Fundamentals of Electrical Substations

FIRST EDITION

BY PRASUN BARUA

ABOUT

Welcome to the "Fundamentals of Electrical Substations"! This book is designed to provide you with a comprehensive understanding of electrical substations, their components, configurations, design considerations, and operational principles. Whether you are a student, engineer, or industry professional, we hope this book serves as a valuable resource to enhance your knowledge and expertise in the field of electrical power systems.

Electrical substations are the backbone of the power grid, facilitating the transformation, transmission, and distribution of electricity from the point of generation to end-users. Their significance in ensuring the reliable, efficient, and safe delivery of electrical power cannot be overstated. As the energy landscape evolves, substations are evolving too, embracing new technologies, integrating renewable energy sources, and adapting to the demands of a changing power sector.

In this book, we have endeavored to provide a comprehensive overview of electrical substations, covering a wide range of topics. We begin with an introduction to substations, exploring their historical background and evolution over time. We then delve into the components and configuration of substations, including transformers, circuit breakers, switchgear, and busbars. Detailed explanations and insights into their operation, ratings, protection mechanisms, and cooling systems are provided.

Design considerations, such as layout, clearances, grounding, and safety protocols, are discussed in detail to highlight the critical aspects that go into planning and constructing substations. We also explore the environmental and aesthetic aspects of substation design, emphasizing the importance of minimizing the environmental impact and integrating substations harmoniously into their surroundings.

The book further delves into substation automation and control systems, discussing the advancements in SCADA systems, remote control technologies, and the integration of smart grid features. We explore the critical role of protection systems in substations and their various applications, including transformer, generator, and transmission line protection. The book also covers the emerging trends and future outlook of substations, touching upon topics such as renewable energy integration, smart grid technologies, and advanced communication and control systems.

Real-world case studies provide practical insights into substation projects, highlighting the challenges, successes, and lessons learned. We also address the importance of regular maintenance, asset management strategies, and diagnostic techniques for equipment condition monitoring, ensuring the optimal performance and longevity of substations.

Throughout the book, we emphasize the significance of safety protocols, regulatory compliance, and environmental responsibility in substations. We aim to equip you with a holistic understanding of the complexities and considerations involved in designing, operating, and maintaining electrical substations.

We hope this book serves as an authoritative and comprehensive guide, allowing you to deepen your knowledge, broaden your perspective, and navigate the dynamic landscape of electrical substations. We encourage you to explore the chapters, engage with the information presented, and apply the insights gained to your professional endeavors.

We express our sincere appreciation to you, the reader, for choosing to embark on this educational journey with us. We hope you find this book insightful, enlightening, and enjoyable. May it inspire your curiosity, enhance your understanding, and empower you to contribute to the advancement of electrical power systems.

TABLE OF CONTENTS

CHAPTER 1: INTRODUCTION TO ELECTRICAL SUBSTATIONS

In the ever-expanding world of electrical power distribution, electrical substations play a vital role in delivering electricity efficiently and reliably to end-users. This chapter serves as an introductory guide to electrical substations, offering an overview of their purpose, historical background, and their critical importance in modern power systems.

Chapter 1 provides a comprehensive introduction to electrical substations, highlighting their crucial role in power distribution systems. It emphasizes the historical context of their development, tracing the evolution from basic local distribution points to the advanced facilities seen in modern power systems. The chapter also emphasizes the importance of substations in ensuring the reliable, efficient, and sustainable delivery of electricity to end-users. By setting the foundation for further exploration, this chapter enables readers to delve into the intricacies of substation components, configurations, design considerations, and maintenance practices in subsequent chapters.

Understanding the Role of Electrical Substations in Power Distribution

Electrical substations play a critical role in power distribution systems by serving as key hubs for the efficient and reliable transfer of electricity from power generation sources to end-

5

users. They act as intermediate points between the high-voltage transmission system and the low-voltage distribution network, ensuring that electricity is delivered at appropriate voltage levels for consumption. To fully grasp the significance of electrical substations in power distribution, it is essential to explore their primary functions and the overall power flow within the grid.

Transformation and Voltage Regulation:

One of the primary functions of electrical substations is to transform electrical energy from one voltage level to another. Power generated at power plants typically operates at high voltages, often ranging from hundreds of kilovolts (kV) to several hundreds of kilovolts (kV). These high voltages are suitable for long-distance transmission due to reduced losses and improved efficiency. However, these high voltages are not suitable for direct consumption by end-users.

Electrical substations serve as transformation points where the voltage is stepped down to lower levels for distribution.

Transmission substations, located at the junctions of high-voltage transmission lines, typically reduce the voltage levels from transmission levels to lower subtransmission or distribution levels (usually in the range of tens of kilovolts). Distribution substations, situated closer to residential, commercial, and industrial areas, further reduce the voltage to levels suitable for local distribution (often in the range of a few kilovolts).

Voltage regulation is another critical function performed by electrical substations. Fluctuations in voltage levels can affect the performance and efficiency of electrical equipment. Substations employ voltage regulation devices such as tap changers in transformers to maintain a steady voltage within predefined limits. This ensures that end-users receive electricity at a consistent and reliable voltage, optimizing the operation of electrical appliances and machinery.

Power Flow Control and Switching Operations:

Electrical substations enable efficient power flow control and switching operations within the power distribution system. They provide the infrastructure necessary to reroute power flows, manage network configuration changes, and handle contingencies such as equipment failures or faults. Switching substations, specifically designed for this purpose, act as interconnection points where power lines can be redirected, isolated, or interconnected based on the desired power flow patterns and network requirements.

Through the use of switches, circuit breakers, and other control devices, substations can isolate faulty sections of the network, thereby minimizing the impact of faults and ensuring continuity of power supply to unaffected areas. This capability enhances the reliability and resilience of power distribution systems, reducing downtime and minimizing disruptions to end-users.

Protection and Fault Management:

Electrical substations are equipped with advanced protection systems designed to detect and mitigate faults within the power distribution system. Faults can occur due to various reasons, such as equipment failures, lightning strikes, or environmental factors. Substation protection systems utilize relays, circuit breakers, and coordination schemes to identify and isolate faults swiftly. By isolating faulted sections and ensuring the rapid restoration of unaffected areas, substations minimize the impact of faults on the overall power distribution network.

Monitoring and Control:

Modern electrical substations incorporate advanced monitoring and control systems that enable real-time supervision and control of various substation components and equipment. Supervisory Control and Data Acquisition (SCADA) systems, along with intelligent electronic devices (IEDs) and communication networks, facilitate remote monitoring, control, and automation of substation operations. This capability enables operators to monitor key parameters, identify potential issues, and take proactive measures to maintain optimal performance and reliability of the substation.

Conclusion:

Understanding the role of electrical substations in power distribution is essential for comprehending the intricate functioning of modern power systems. These substations serve as vital links between power generation sources and end-users, transforming and regulating voltage levels, managing power flows, protecting the network from faults, and enabling efficient control and monitoring. By performing these functions effectively, electrical substations contribute to the reliable and uninterrupted supply of electricity to homes, businesses, and industries, supporting economic growth and enhancing the quality of life for millions of people.

Historical Background and Evolution of Electrical Substations

The development and evolution of electrical substations can be traced back to the late 19th century when the demand for electricity began to surge due to the rapid advancements in electrical technology. The establishment of electrical power distribution systems necessitated the creation of intermediate points between power generation sources and end-users to ensure efficient and reliable electricity supply. The historical background of electrical substations showcases a progression from rudimentary local distribution points to the sophisticated facilities seen in modern power systems.

Early Electrical Distribution Systems:

In the early days of electricity, power distribution systems were relatively simple, primarily serving localized areas. Direct current (DC) systems were commonly used, with generators supplying electricity to nearby consumers through copper or iron conductors. These systems often operated at low voltages, limiting their transmission distances. Consequently, substations were not widely employed during this period, as the focus was on establishing small-scale electrical networks to cater to specific geographical areas.

The Advent of Alternating Current (AC) and Power Transmission:

The discovery and development of alternating current (AC) by pioneers like Nikola Tesla and George Westinghouse revolutionized power distribution. AC allowed for efficient transmission of electrical power over long distances, enabling the establishment of larger-scale power generation plants located far away from urban centers. The use of transformers facilitated the conversion of voltage levels, making it feasible to transmit electricity at higher voltages and then step it down at substations closer to the consumers.

Introduction of Electrical Substations:

As power systems grew in size and complexity, the need for intermediate points between power generation and consumption became apparent. These intermediate points, known as electrical substations, served as essential junctions in the power grid. Early substations were modest in terms of equipment and functionality, primarily consisting of transformers and switches for local distribution. They were often located in close proximity to power plants or in centralized areas to facilitate the distribution of electricity to nearby consumers.

Expansion of Power Grids and Transmission Substations:

With the expanding power generation capacities and the development of inter-regional power grids, transmission substations became crucial components of the electrical infrastructure. These substations were strategically positioned at the intersections of high-voltage transmission lines, allowing power to be stepped down from transmission levels to lower subtransmission or distribution levels. Transmission substations often housed multiple transformers and switchgear to facilitate the efficient transfer of power across extensive networks.

Advancements in Substation Design and Technology:

Throughout the 20th century, significant advancements in substation design and technology took place, driven by the growing demand for electricity and the need for enhanced reliability and efficiency. Substation components and equipment evolved to incorporate more sophisticated features and capabilities. Developments included the introduction of advanced protection systems, switchgear with improved fault-clearing capabilities, and monitoring and control systems for remote operation and maintenance.

Automation and Digitalization of Substations:

In recent decades, the automation and digitalization of electrical substations have become prominent trends. The integration of intelligent electronic devices (IEDs), microprocessor-based relays, and communication networks has revolutionized substation control and monitoring. Supervisory Control and Data Acquisition (SCADA) systems enable operators to remotely supervise and manage substations, improving operational efficiency, reliability, and maintenance practices. Additionally, the introduction of substation automation systems and smart grid technologies has enhanced grid resilience, flexibility, and the integration of renewable energy sources.

Conclusion:

The historical background of electrical substations showcases the evolution of these crucial components in power distribution systems. From their humble beginnings as basic local distribution points, substations have evolved into sophisticated facilities that enable the efficient and reliable transfer of electricity across extensive power grids. Advances in substation design, technology, and automation have significantly enhanced the performance, safety, and operational capabilities of these critical nodes in the power infrastructure. As the demand for electricity continues to grow and new energy sources emerge, electrical substations will continue to evolve, adapting to the changing needs of the power industry.

Importance of Substations in Modern Power Systems

Substations play a crucial role in modern power systems, serving as vital components that ensure the efficient, reliable, and safe distribution of electricity from power generation sources to end-users. The importance of substations can be understood by examining their key contributions to the overall functioning and stability of power systems.

Voltage Transformation and Regulation:

One of the primary functions of substations is to transform electrical energy from one voltage level to another. Power generated at power plants typically operates at high voltages, enabling efficient long-distance transmission. However, these high voltages are not suitable for direct consumption by end-users. Substations step down the voltage levels to more manageable levels for distribution, ensuring that electricity is delivered to consumers at appropriate voltages. This voltage transformation and regulation provided by substations enable the safe and efficient utilization of electrical energy across various sectors.

Power Flow Control and Network Configuration:

Substations play a pivotal role in managing power flow and network configuration within power systems. By strategically locating substations at key junctions of transmission and distribution lines, power flows can be controlled and redirected based on demand and network conditions. Substations provide switching operations that enable the reconfiguration of power networks, allowing for load balancing, optimization of power flow routes, and the ability to isolate faulted sections to minimize disruptions. This capability enhances the stability, reliability, and flexibility of power systems, ensuring that electricity is delivered efficiently to consumers.

Fault Detection and Mitigation:

Electrical substations are equipped with advanced protection systems designed to detect and mitigate faults within power systems. Faults can occur due to equipment failures, lightning strikes, or other unforeseen events. Substation protection schemes, consisting of relays, circuit breakers, and coordination schemes, are implemented to identify faults accurately and isolate affected sections. By swiftly isolating faulted areas and restoring unaffected areas, substations mitigate the impact of faults, minimize downtime, and enhance the reliability and quality of power supply to consumers.

Monitoring, Control, and Automation:

Modern substations incorporate advanced monitoring, control, and automation systems that enable real-time supervision and operation of substation components. Supervisory Control and Data Acquisition (SCADA) systems, along with intelligent electronic devices (IEDs) and communication networks, facilitate remote monitoring, control, and automation of substation operations. This capability allows operators to monitor critical parameters, detect abnormalities, and respond promptly to maintain optimal performance. Automation technologies enable the efficient management of substations, reducing human intervention, enhancing operational efficiency, and improving response times during contingencies.

Integration of Renewable Energy Sources:

As the integration of renewable energy sources, such as solar and wind power, increases, substations play a crucial role in enabling their seamless integration into the power grid. Substations facilitate the connection and synchronization of renewable energy sources with the existing power infrastructure, ensuring their efficient injection into the grid. Substations also contribute to managing the intermittency and variability of renewable energy sources through advanced control and monitoring capabilities, allowing for better grid stability and improved utilization of clean energy resources.

Grid Resilience and Energy Security:

Substations contribute to the overall resilience and security of power grids. By enabling load balancing, rerouting of power flows, and quick fault detection and isolation, substations help minimize the impact of disruptions and enhance grid reliability. Additionally, substations provide essential infrastructure for emergency response and restoration activities during power outages, ensuring that power can be restored efficiently and swiftly.

Conclusion:

The importance of substations in modern power systems cannot be overstated. They serve as critical links between power generation sources and end-users, ensuring efficient voltage transformation, power flow control, fault detection, and mitigation. Substations also facilitate the integration of renewable energy sources, contribute to grid resilience, and enable effective monitoring, control, and automation of power systems. By fulfilling these roles, substations enhance the reliability, stability, and efficiency of power distribution, supporting economic growth, and meeting the increasing energy demands of society.

CHAPTER 2: COMPONENTS AND CONFIGURATION OF ELECTRICAL SUBSTATIONS

Chapter 2 delves into the various components and configurations of electrical substations, providing a comprehensive understanding of the essential elements that make up these critical power system assets. By exploring the functionalities, characteristics, and interconnections of different substation components, readers will gain insight into the intricate design and configuration considerations involved in constructing substations.

Chapter 2 provides a comprehensive overview of the various components and configurations within electrical substations. By exploring transformers, circuit breakers, switchgear, busbars, protective relays, auxiliary equipment, and substation configurations, readers will gain a deeper understanding of the intricacies involved in designing and constructing these critical power system assets. This knowledge will enable readers to make informed decisions regarding the selection, configuration, and maintenance of substation components, ensuring the efficient and reliable operation of electrical substations within power systems.

Overview of Substation Components: Transformers, Circuit Breakers, Switchgear, Busbars, and More

Substations are complex electrical facilities that consist of various components working together to ensure the efficient transmission, distribution, and control of electric power. This section provides an overview of some of the essential substation components, including transformers, circuit breakers, switchgear, busbars, and other auxiliary equipment.

Transformers:

Transformers are key components in electrical substations that facilitate the transformation of electrical energy from one voltage level to another. They play a critical role in stepping up or stepping down the voltage levels to match the requirements of power transmission and distribution. Transformers are available in various types, including power transformers, distribution transformers, and auto-transformers. They are rated based on voltage ratios, power ratings, and impedance. Cooling systems, such as oil-filled or dry-type, ensure efficient heat dissipation. Transformers are also equipped with protective devices like Buchholz relays and temperature monitoring systems to ensure their safe operation.

Circuit Breakers:

Circuit breakers are devices designed to interrupt or disconnect electrical circuits during abnormal conditions, such as short circuits or overloads. They play a crucial role in protecting substation equipment and maintaining the overall stability of the power system. Different types of circuit breakers are used in substations, including air blast, vacuum, SF6 (sulfur hexafluoride), and oil-filled circuit breakers. Circuit breakers are rated based on their current and voltage capacities, as well as their interrupting capacity. Protective relays are often coordinated with circuit breakers to detect faults and provide selective tripping or reclosing capabilities.

Switchgear:

Switchgear refers to the combination of electrical switches, circuit breakers, and other control devices used for controlling, protecting, and isolating electrical equipment within a substation. Switchgear can be categorized into different types, such as metal-clad switchgear, metal-enclosed switchgear, and gas-insulated switchgear (GIS). It provides a means for safely isolating faulty sections, redirecting power flows, and controlling the overall operation of the substation. Switchgear is an integral part of fault protection and ensures the safe and reliable operation of substation equipment.

Busbars:

Busbars are conductive bars that act as a common electrical connection or distribution point for multiple circuits within a substation. They provide a low-impedance path for the flow of electric current and facilitate the distribution of power to various substation components. Busbars are designed to handle high currents and are made of materials with excellent electrical conductivity, such as copper or aluminum. Different configurations of busbars, such as single bus, double bus, ring bus, or breaker and a half, are used depending on the specific substation requirements and desired operational flexibility.

Protective Relays:

Protective relays are devices that detect abnormal electrical conditions, such as overcurrent, under/over voltage, or differential current, and initiate appropriate actions to isolate the faulty sections of the power system. Protective relays play a crucial role in identifying faults, ensuring selective tripping of circuit breakers, and coordinating protection schemes to maintain the stability and reliability of the substation. They are designed to respond quickly and accurately to abnormal electrical conditions, providing a high level of system protection.

Auxiliary Equipment:

Auxiliary equipment in substations includes various components that support the overall functionality and safety of the substation. This equipment may include capacitors and reactors for power factor correction and reactive power control, battery and charger systems for backup power supply, grounding systems for safety and effective fault current dissipation, lightning protection systems to protect against surges, and fire detection and suppression systems to mitigate fire hazards within the substation.

Conclusion:

Understanding the various components within electrical substations is crucial for comprehending their overall functioning and ensuring the efficient and reliable operation of power systems. Transformers, circuit breakers, switchgear, busbars, protective relays, and auxiliary equipment all play vital roles in power transmission, distribution, protection, and control. By working in tandem, these components enable the safe and efficient delivery of electric power from generation sources to end-users, contributing to the stability and reliability of electrical substations and the overall power grid.

Different Types of Substations: Transmission, Distribution, and Switching Substations

Electrical substations can be classified into several types based on their functionality, voltage levels, and specific roles within the power system. The primary types of substations are transmission substations, distribution substations, and switching substations. Understanding the characteristics and functions of each type is crucial for comprehending the overall power system infrastructure.

Transmission Substations:

Transmission substations are a critical component of the power grid responsible for the transmission of electricity over long distances at high voltages. These substations are typically located at the junctions of high-voltage transmission lines and act as interfaces between the power generation sources, such as power plants, and the distribution network. The main functions of transmission substations include:

- Stepping up or stepping down voltage levels to facilitate efficient long-distance transmission. They receive power from generating stations at lower voltages and elevate the voltage to transmission levels, which can range from hundreds of kilovolts (kV) to several hundreds of kilovolts (kV).
- Ensuring the reliability and stability of the transmission system by providing reactive power compensation and voltage control.
- Providing switching capabilities to enable rerouting of power flows and reconfiguration of transmission networks during maintenance or contingencies.

Transmission substations usually house large power transformers, circuit breakers, and switchgear suitable for high-voltage operations. They may also include various auxiliary systems for monitoring, control, and protection. Due to the higher voltages involved, transmission substations have stricter safety protocols and typically have restricted access.

Distribution Substations:

Distribution substations are located closer to the end-users and are responsible for stepping down the voltage from transmission levels to levels suitable for local distribution. They receive power from transmission substations at higher voltages and reduce the voltage to distribution levels, which can range from tens of kilovolts (kV) to a few kilovolts (kV). The main functions of distribution substations include:

- Further reducing the voltage level for distribution to residential, commercial, and industrial areas.
- Ensuring a reliable and stable power supply to end-users by incorporating protective devices, such as circuit breakers and protective relays, for fault detection and isolation.
- Enabling sectionalization and the creation of multiple feeders to distribute power to different areas and customers.
- Providing transformer capacity to match the load requirements of the distribution system.

Distribution substations typically contain power transformers, circuit breakers, switchgear, and protective devices suitable for lower voltage levels. These substations may also include capacitor banks for power factor correction and voltage regulation, as well as monitoring and control systems for efficient operation and maintenance.

Switching Substations:

Switching substations, also known as switching stations or distribution switching stations, primarily serve as interconnection points or switching nodes within the distribution network. Their main functions include:

- Providing switching capabilities to enable rerouting of power flows, isolation of faulted sections, and coordination of power supply during maintenance or contingencies.

- Facilitating load management by allowing the redirection of power to balance loads and optimize network operation.
- Offering flexibility in the distribution system by providing alternative routes and redundancy to enhance reliability.

Switching substations typically consist of switchgear, circuit breakers, and protective devices designed for distribution voltage levels. These substations may be strategically located along the distribution network to ensure efficient power flow control and network reconfiguration.

It's important to note that the classification of substations is not always rigid, and there can be variations or hybrid configurations depending on specific regional or operational requirements. Additionally, with the evolution of power systems and the integration of renewable energy sources, substations are also adapting to accommodate emerging technologies and facilitate the seamless integration of clean energy into the grid.

Conclusion:

Transmission substations, distribution substations, and switching substations are integral components of the power system infrastructure. Each type serves specific functions within the overall power distribution network. Transmission substations focus on high-voltage transmission, distribution substations deliver electricity at appropriate voltage levels to end-users, and switching substations enable power flow control and network reconfiguration. By understanding the roles and characteristics of these different types of substations, power system operators can effectively plan, design, and operate the electrical infrastructure, ensuring the efficient and reliable delivery of electricity to consumers.

Substation Layouts and Configurations: Single Bus, Double Bus, Ring Bus, Breaker and a Half, and Others

The layout and configuration of electrical substations are crucial considerations in the design and operation of power systems. Different substation layouts and configurations are employed based on factors such as the voltage level, system requirements, reliability, operational flexibility, and maintenance considerations. This section provides an elaborate exploration of various substation layouts and configurations, including single bus, double bus, ring bus, breaker and a half, and others.

Single Bus Substation Configuration:

The single bus configuration is the simplest and most basic layout used in substations. In this configuration, all incoming and outgoing transmission or distribution lines are connected to a single busbar. The main features of the single bus configuration include:

- Minimal complexity and lower construction costs due to a reduced number of components.
- Limited redundancy and operational flexibility, as any maintenance or fault on the single busbar can lead to a complete outage of the substation.
- Suitable for smaller substations or situations where redundancy is not a critical requirement.

Double Bus Substation Configuration:

The double bus configuration enhances the operational flexibility and reliability of the substation compared to the single bus configuration. In this layout, two busbars are utilized, with each busbar capable of handling the entire load of the substation independently. The key features of the double bus configuration include:

- Increased redundancy, allowing for maintenance or fault isolation without affecting the overall operation of the substation.
- Flexibility to transfer loads between busbars during maintenance or emergencies.
- Complex switching operations and higher construction costs compared to the single bus configuration.
- Suitable for substations where reliability, operational flexibility, and ease of maintenance are important factors.

Ring Bus Substation Configuration:

The ring bus configuration is designed to further enhance the reliability and flexibility of substations. In this layout, multiple busbars are interconnected in a ring-like arrangement. The main features of the ring bus configuration include:

- High redundancy, as any fault or maintenance on one section of the ring does not disrupt the overall operation of the substation.
- Enhanced reliability and fault tolerance, allowing for automatic reconfiguration of power flows in case of faults.
- Complex switching and protection schemes to maintain the integrity of the ring during faults.
- Suitable for critical substations where uninterrupted power supply is crucial, such as hospitals, data centers, and industrial facilities.

Breaker and a Half Substation Configuration:

The breaker and a half configuration combines elements of the double bus and ring bus configurations. It features two main busbars and additional bus sections connected by circuit breakers. The key features of the breaker and a half configuration include:

- Increased flexibility and redundancy compared to the double bus configuration.
- Allows for selective isolation of faulty sections while maintaining power supply to the rest of the substation.
- Complex switching schemes and higher construction costs compared to other configurations.
- Suitable for substations where high reliability, operational flexibility, and maintenance ease are critical.

Other Substation Configurations:

In addition to the commonly used layouts mentioned above, there are other specialized substation configurations employed in specific situations. These include:

- Transfer Bus Substation: Designed to facilitate the transfer of loads from one transmission line to another without disruption.
- Mesh Bus Substation: Utilizes multiple interconnected busbars to enhance redundancy and provide alternate power supply paths.
- Hybrid Substation: Combines different configurations and technologies to optimize performance and meet specific requirements.

Each substation layout and configuration has its advantages and considerations. Factors such as system requirements, load characteristics, fault tolerance, operational flexibility, maintenance requirements, and cost considerations play a role in selecting the most suitable configuration for a given substation.

Conclusion:

Substation layouts and configurations play a significant role in the reliability, flexibility, and efficiency of power systems. Whether it is the simplicity of the single bus configuration, the redundancy of the double bus or ring bus configurations, or the flexibility of the breaker and a half configuration, each layout serves specific purposes based on operational requirements and system characteristics. By carefully considering the advantages and trade-offs associated with different substation layouts and configurations, engineers and system operators can design and operate substations that meet the demands of a reliable and efficient power distribution network.

CHAPTER 3: SUBSTATION EQUIPMENT

Chapter 3 focuses on the various equipment used in electrical substations. These equipment components are essential for the proper functioning and operation of substations. This chapter provides an in-depth exploration of key substation equipment, including transformers, circuit breakers, switchgear, protective relays, control systems, and auxiliary equipment.

Chapter 3 provides a comprehensive overview of the equipment used in electrical substations. Transformers, circuit breakers, switchgear, protective relays, control systems, and auxiliary equipment are all critical for the reliable and efficient operation of substations. Understanding the functions, characteristics, and interconnections of these equipment components is essential for designing, operating, and maintaining substations. By delving into the intricacies of substation equipment, this chapter equips readers with the knowledge necessary to ensure the optimal performance and reliability of electrical substations in power systems.

Detailed Exploration of Transformers: Types, Ratings, Cooling Systems, and Protection

Transformers are vital components of electrical substations, responsible for the efficient and reliable transformation of electrical energy from one voltage level to another. This section provides a detailed exploration of transformers, including their types, ratings, cooling systems, and protection mechanisms.

Types of Transformers:

Power Transformers: Power transformers are commonly used in substations for high voltage (HV) to medium voltage (MV) or medium voltage to low voltage (LV) applications. They are designed to handle large power capacities and operate at high voltages. Power transformers are typically used for voltage step-up or step-down purposes in transmission and distribution systems.

Distribution Transformers: Distribution transformers are primarily used for step-down purposes in distribution networks. They transform the voltage from medium voltage to low voltage levels suitable for consumer utilization. Distribution transformers are typically located closer to the end-users and are available in various sizes and configurations.

Auto-Transformers: Auto-transformers differ from conventional transformers in that they share a common winding between the primary and secondary sides. This design allows them to be more compact and efficient compared to traditional transformers. Auto-transformers are often used in applications where significant voltage ratios are not required.

Transformer Ratings and Specifications:

Voltage Ratings: Transformers are designed to operate at specific voltage levels. The primary voltage rating refers to the voltage at the input (high voltage side), while the secondary voltage rating represents the voltage at the output (low voltage side). Transformers may have multiple taps on the winding to allow for minor adjustments in voltage levels.

Power Ratings: Power transformers are rated based on their power handling capacities, expressed in volt-amperes (VA) or kilovolt-amperes (kVA). The power rating indicates the maximum power that the transformer can transfer without exceeding its thermal and electrical limits.

Impedance: Transformer impedance refers to the internal resistance to the flow of current. It is specified as a percentage and impacts the voltage regulation and fault current capability of the transformer.

Cooling Systems:

Transformers generate heat during operation due to electrical losses, and efficient cooling is necessary to maintain their temperature within safe limits. Various cooling systems are employed in transformers, including:

1. **Oil-Filled Cooling:** Oil-filled transformers use mineral oil or synthetic ester as a cooling medium. The oil absorbs heat generated by the transformer's windings and dissipates it through convection and radiation. Radiators and fans may be used to enhance cooling.
2. **Dry-Type Cooling:** Dry-type transformers use air as the cooling medium. They are insulated with epoxy resin or cast resin and do not require the use of oil. Air circulation is achieved through natural convection or forced-air cooling using fans.
3. **Forced-Air Cooling:** Forced-air cooling is commonly employed in large power transformers. Fans or blowers force air through cooling fins or radiators to enhance heat dissipation.

Transformer Protection and Monitoring:

To ensure the safe and reliable operation of transformers, various protection and monitoring measures are implemented:

1. **Buchholz Relay:** Buchholz relay is a protective device installed in oil-filled transformers. It detects and responds to internal faults, such as electrical faults and gas accumulation, by initiating an alarm or tripping the transformer. It provides early warning signs of developing faults and helps prevent catastrophic failures.

2. **Temperature Monitoring:** Transformers are equipped with temperature sensors placed in critical locations to monitor the temperature rise. Abnormal temperature increases can indicate overloading, faults, or insufficient cooling. Temperature monitoring allows for timely intervention and preventive maintenance.

3. **Differential Protection:** Differential protection is employed to detect internal faults within the transformer windings. It compares the currents entering and exiting the transformer to identify any imbalances, indicating a fault condition. Differential protection initiates a trip signal to isolate the faulty section and prevent further damage.

4. **Ground-Fault Protection:** Ground-fault protection safeguards transformers from insulation failures or short circuits to ground. It detects the flow of ground fault current and initiates a protective action, such as tripping the circuit breaker, to isolate the fault and prevent damage to the transformer.

5. **Overcurrent Protection:** Overcurrent protection devices, such as protective relays, ensure that the transformer is not subjected to excessive current during abnormal conditions.

Overcurrent protection prevents overheating and damage to the transformer's windings and insulation.

Conclusion:

Transformers are critical components of electrical substations, responsible for voltage transformation and ensuring the efficient and reliable transmission and distribution of electrical power. Understanding the different types of transformers, their ratings, cooling systems, and protection mechanisms is essential for the design, operation, and maintenance of electrical substations. By implementing appropriate transformer configurations, ratings, cooling systems, and protection measures, power system operators can ensure the optimal performance and longevity of transformers, contributing to the overall stability and reliability of the power grid.

Circuit breakers: types, operation, and protection mechanisms

Circuit breakers are vital components in electrical substations, designed to protect electrical systems from faults and overloads. They play a crucial role in interrupting fault currents, isolating faulty sections, and maintaining the integrity and stability of the power grid. This section provides a comprehensive exploration of circuit breakers, including their types, operation, and protection mechanisms.

Types of Circuit Breakers:

1. **Air Blast Circuit Breakers:** Air blast circuit breakers use compressed air to extinguish the arc that forms when the circuit breaker interrupts the fault current. The high-pressure air blast cools and clears the arc, ensuring rapid interruption. Air blast circuit breakers are commonly used in high-voltage applications.
2. **Vacuum Circuit Breakers:** Vacuum circuit breakers utilize a vacuum as the arc-quenching medium. When

the circuit breaker operates, contacts within the vacuum interrupt the current flow, extinguishing the arc. Vacuum circuit breakers are compact, reliable, and commonly used in medium-voltage applications.

3. **SF6 (Sulfur Hexafluoride) Circuit Breakers:** SF6 circuit breakers employ SF6 gas as the arc-quenching medium. SF6 gas provides excellent insulation and arc-extinguishing properties. These circuit breakers are commonly used in high-voltage applications due to their compact size and high interrupting capability.

4. **Oil-Filled Circuit Breakers:** Oil-filled circuit breakers use oil as the arc-quenching medium. The oil provides insulation and cooling properties. Oil-filled circuit breakers were commonly used in the past but are gradually being replaced by more environmentally friendly alternatives.

Operation of Circuit Breakers:

Circuit breakers operate through a series of well-defined stages:

1. **Closed Position:** In the closed position, the circuit breaker contacts are connected, allowing the current to flow through the system.
2. **Trip Operation:** When a fault or abnormal condition occurs, the protective relay senses the fault and sends a trip signal to the circuit breaker. The trip signal initiates the opening operation.
3. **Arc Extinction:** As the circuit breaker opens, an arc is formed between the separating contacts. The arc needs to be extinguished to interrupt the current flow. The circuit breaker's design and the arc-quenching medium determine the method used to extinguish the arc.

4. **Contact Separation:** After the arc is extinguished, the contacts of the circuit breaker are separated, ensuring complete isolation of the faulted section.
5. **Arc Clearance:** Once the contacts are separated, the arc needs to be cleared. This process can involve various mechanisms, such as mechanical devices, blast valves, or cooling systems, depending on the circuit breaker type.
6. **Re-Closing:** In some cases, after the fault is cleared, circuit breakers can be re-closed to restore power to the system. Re-closing can be done manually or automatically, depending on the application and system requirements.

Protection Mechanisms:

Circuit breakers incorporate various protection mechanisms to ensure safe and reliable operation:

1. **Overcurrent Protection:** Overcurrent protection devices, such as protective relays, sense excessive current flowing through the circuit. They initiate a trip signal to the circuit breaker to interrupt the current and prevent equipment damage or hazardous conditions.
2. **Short-Circuit Protection:** Short-circuit protection detects fault currents caused by short circuits and initiates the tripping operation to isolate the faulted section. Short-circuit protection is crucial for preventing equipment damage and minimizing system downtime.
3. **Ground-Fault Protection:** Ground-fault protection detects faults that occur when an electrical circuit comes in contact with ground. It senses the flow of ground fault current and initiates a protective action to

isolate the fault and prevent damage to equipment and electrical systems.

4. **Overload Protection:** Overload protection devices monitor the current flowing through the circuit and detect prolonged excessive currents. When the current exceeds the predetermined threshold, the overload protection initiates a trip signal to the circuit breaker to prevent overheating and damage to the system.

Conclusion:

Circuit breakers are critical components in electrical substations, providing fault protection and maintaining the stability of power systems. Understanding the different types of circuit breakers, their operation mechanisms, and protection functionalities is essential for designing, operating, and maintaining substations. By implementing appropriate circuit breaker types and incorporating effective protection mechanisms, power system operators can ensure the safe and reliable operation of electrical systems, protecting equipment and minimizing disruptions in the power grid.

Switchgear and Protective Relays: Functions, Coordination, and Safety Features

Switchgear and protective relays are integral components in electrical substations, responsible for the control, protection, and isolation of electrical equipment. This section provides an elaborate exploration of switchgear and protective relays, including their functions, coordination, and safety features.

Switchgear Functions:

Switchgear refers to the combination of electrical switches, circuit breakers, and control devices used to control, protect, and isolate electrical equipment within a substation. The functions of switchgear include:

1. **Control:** Switchgear allows operators to control the flow of electrical power within a substation. It provides switching operations to enable the connection or disconnection of electrical equipment, facilitating safe and efficient operation.
2. **Protection:** Switchgear incorporates protective devices, such as circuit breakers and fuses, to detect and interrupt abnormal electrical conditions, such as overcurrent, short circuits, and faults. These protective devices ensure the safe operation of electrical systems and prevent damage to equipment and personnel.
3. **Isolation:** Switchgear allows for the isolation of faulty sections or equipment within a substation. By isolating faulty components, switchgear minimizes the impact of faults and prevents them from spreading to other parts of the system, enhancing safety and reliability.

Protective Relay Functions:

Protective relays work in coordination with switchgear to detect abnormal electrical conditions and initiate appropriate actions to protect the system. The functions of protective relays include:

1. **Fault Detection:** Protective relays continuously monitor electrical parameters, such as current, voltage, and frequency, to detect abnormal conditions indicative of faults. They sense deviations from normal operating conditions and initiate appropriate responses to isolate faulty sections.
2. **Discrimination:** Protective relays coordinate with each other to selectively isolate the faulted section while minimizing the impact on the rest of the system. Through careful coordination and grading, protective relays ensure that only the circuit breaker closest to the fault is tripped, preventing unnecessary interruptions in the power supply.

3. **Timing and Coordination:** Protective relays operate based on predefined time-current characteristics. The coordination of protective relays involves setting specific time-delay characteristics to allow downstream protective devices to clear faults before upstream devices. This coordination ensures that the fault is isolated quickly and selectively, reducing downtime and maintaining system stability.

Safety Features:

Switchgear and protective relays incorporate several safety features to enhance the safety of electrical systems:

1. **Mechanical Interlocks:** Switchgear is equipped with mechanical interlocks that prevent incorrect or unsafe operations. These interlocks ensure that specific operations, such as closing a circuit breaker when another circuit breaker is already closed, are physically inhibited to avoid hazardous conditions.

2. **Safety Interlocks:** Safety interlocks are electrical or mechanical devices that prevent unsafe conditions or operations. For example, safety interlocks can prevent access to live parts or inhibit operations that can lead to hazardous situations, ensuring the safety of personnel working with switchgear.

3. **Arc-Flash Protection:** Switchgear may include arc-flash protection devices, such as arc-resistant enclosures or arc-flash detection systems. These safety measures are designed to detect and mitigate arc-flash incidents, reducing the risk of injuries and equipment damage due to arc faults.

4. **Grounding and Bonding:** Proper grounding and bonding systems are essential safety features in switchgear

installations. They ensure the dissipation of fault currents, provide a safe path for fault current flow, and minimize the risk of electric shock to personnel.

5. **Safety Labels and Signage:** Switchgear is typically labeled with safety information, warning signs, and operational instructions. These labels and signs provide important guidance to personnel, ensuring they understand the risks associated with the equipment and how to safely operate and maintain it.

Coordination between Switchgear and Protective Relays:

Switchgear and protective relays work together in a coordinated manner to ensure proper protection and operation of electrical systems. Coordination involves:

1. **Selectivity:** The coordination of protective devices, such as protective relays and circuit breakers, ensures selective tripping. This means that only the device closest to the fault operates to isolate the faulted section, minimizing system disruptions and downtime.

2. **Grading:** Grading involves setting the time-current characteristics of protective relays to create a coordinated response during fault conditions. The settings are established to allow downstream protective devices to operate before upstream devices, ensuring that faults are cleared from the system in a sequential and coordinated manner.

Conclusion:

Switchgear and protective relays are crucial components in electrical substations, providing control, protection, and isolation of electrical equipment. Switchgear enables safe and efficient operation, while protective relays detect faults and initiate appropriate actions for system protection. Through careful coordination and the incorporation of safety features, switchgear and protective relays enhance the safety, reliability, and efficiency of electrical systems, protecting equipment and personnel from electrical hazards and minimizing disruptions in the power grid.

Capacitors, reactors, and other auxiliary equipment

Capacitors, reactors, and other auxiliary equipment are important components in electrical substations, serving various functions to ensure efficient and reliable operation of power systems. This section provides an elaborate exploration of capacitors, reactors, and other auxiliary equipment, including their roles, applications, and benefits.

Capacitors:

Capacitors are devices used to store and release electrical energy. In substations, capacitors are primarily used for power factor correction, voltage regulation, and reactive power compensation. The key aspects of capacitors in substations include:

1. **Power Factor Correction:** Capacitors improve the power factor by offsetting the reactive power component in electrical systems. Power factor correction minimizes energy losses, reduces voltage drops, and increases the overall efficiency of the system. Capacitor banks are often connected in parallel with inductive loads to enhance power factor.
2. **Voltage Regulation:** Capacitors help regulate voltage levels by absorbing and releasing reactive power. During low-load periods, capacitors supply reactive power, raising the

voltage. During high-load periods, capacitors absorb excess reactive power, thereby stabilizing the voltage.

3. **Reactive Power Compensation:** Capacitors compensate for the reactive power demand of inductive loads, improving the system's voltage stability and reducing the burden on power generation sources. They help reduce line losses and improve transmission and distribution efficiency.

Capacitors are available in various configurations, including fixed capacitor banks and switched capacitor banks with multiple steps for flexible reactive power compensation. Capacitors may be installed at distribution substations, industrial facilities, and commercial buildings to improve power factor and enhance system performance.

Reactors:

Reactors, also known as inductors or chokes, are passive electrical components used to limit current flow, provide impedance, and mitigate transient disturbances. Key aspects of reactors in substations include:

1. **Current Limiting:** Reactors are used to limit fault currents and control the flow of current in electrical systems. By introducing impedance, reactors reduce short-circuit currents, protecting equipment from damage and preventing disruptions in the power grid.

2. **Harmonic Filtering:** Reactors mitigate harmonic distortion caused by non-linear loads, such as variable frequency drives, computers, and lighting systems. They provide reactive impedance to filter out harmonic frequencies, maintaining the quality of the power supply and preventing adverse effects on equipment and system performance.

3. **Voltage Regulation:** Reactors can be used for voltage regulation by controlling voltage rise or drop. By introducing reactance, reactors help regulate the voltage levels and ensure the proper functioning of equipment within specified voltage limits.

Reactors are available in various types, including air-core reactors, iron-core reactors, and line reactors. They are commonly used in transmission and distribution substations, as well as industrial and commercial facilities, to control current, limit fault levels, and improve system stability.

Battery and Charger Systems:

Battery and charger systems provide backup power supply in substations, ensuring continuous operation during power outages or when the main power source fails. Key aspects of battery and charger systems include:

1. **Uninterrupted Power Supply:** Battery systems, consisting of rechargeable batteries, provide backup power to critical substation equipment, such as control systems, protective relays, and communication systems. They ensure the continuous operation of essential functions during power disruptions.

2. **Charging and Maintenance:** Charger systems are responsible for maintaining the battery's charge, ensuring its readiness for backup power. They monitor the battery's state of charge, voltage, and temperature, providing the necessary charging current to keep the battery in optimal condition.

Battery and charger systems are typically installed in substations where uninterrupted power supply is critical, such as critical infrastructure facilities, data centers, and hospitals. Regular maintenance and testing are necessary to ensure the reliability and performance of these systems.

Grounding Systems:

Grounding systems play a crucial role in substations by providing a safe path for fault currents and ensuring the effective dissipation of electrical energy. Key aspects of grounding systems include:

1. **Safety:** Grounding systems protect personnel and equipment by minimizing the risk of electrical shocks. They provide a low-resistance path for fault currents to flow, facilitating the quick operation of protective devices to isolate faults.
2. **Fault Current Dissipation:** Grounding systems effectively dissipate fault currents, reducing the risk of damage to equipment and minimizing the impact of faults on the overall power system. Proper grounding helps limit voltage rise during faults and ensures the stability and integrity of the electrical system.

Grounding systems include various components, such as ground rods, conductors, grounding grids, and grounding electrodes. They are designed to comply with safety standards and regulations to provide effective grounding and fault current dissipation.

Lightning Protection Systems:

Lightning protection systems safeguard substations against the damaging effects of lightning strikes. Key aspects of lightning protection systems include:

1. **Surge Arresters:** Surge arresters, also known as lightning arresters, are installed at various points in the substation to divert lightning-induced surges to the ground, protecting equipment from voltage transients caused by lightning strikes.

2. **Grounding:** Proper grounding of the substation and equipment is essential for effective lightning protection. Grounding systems ensure the dissipation of lightning currents, preventing damage to equipment and maintaining the safety of personnel.

Lightning protection systems typically include surge arresters, grounding systems, shielding, and proper bonding of metallic structures. They are designed to minimize the impact of lightning-induced surges and safeguard substation equipment.

Conclusion:

Capacitors, reactors, and other auxiliary equipment play crucial roles in electrical substations, enhancing system performance, providing reactive power compensation, mitigating transients, and ensuring reliable and safe operation. Capacitors facilitate power factor correction and voltage regulation, reactors control current flow and mitigate harmonics, and auxiliary equipment like battery systems, grounding systems, and lightning protection systems provide backup power, safety, and protection against external disturbances. By incorporating these auxiliary equipment components, substations can achieve improved power quality, enhanced system stability, and increased reliability, contributing to the efficient and uninterrupted supply of electricity to consumers.

CHAPTER 4: SUBSTATION DESIGN AND PLANNING

Chapter 4 focuses on the design and planning considerations for electrical substations. Proper design and planning are crucial for the efficient, reliable, and safe operation of substations. This chapter provides an in-depth exploration of substation design principles, site selection, layout considerations, equipment arrangement, and safety aspects.

Chapter 4 emphasizes the importance of proper design and planning in the construction and operation of electrical substations. By following design principles, selecting suitable sites, considering layout and equipment arrangement, and incorporating safety and control systems, substations can be optimized for efficiency, reliability, and safety. A well-designed and properly planned substation ensures the secure and uninterrupted flow of electrical power, minimizes downtime, and provides a stable and resilient infrastructure for the power system.

Site Selection Criteria for Substations

The site selection process for electrical substations is a critical step in the design and planning phase. The chosen site should meet various criteria to ensure the efficient, reliable, and safe operation of the substation. This section provides an elaborate exploration of the site selection criteria for substations.

Proximity to Power Sources and Load Centers:

The substation should be located in close proximity to power sources, such as generating stations or transmission lines, to minimize transmission losses and ensure efficient power transfer. It should also be situated near load centers to reduce transmission and distribution distances, thereby minimizing voltage drop and improving system reliability.

Accessibility:

The site should be easily accessible for construction, operation, and maintenance activities. Good road connections and proximity to transportation networks are crucial to facilitate the delivery of equipment, materials, and personnel. Adequate space should be available for construction staging, temporary storage, and installation of heavy machinery.

Environmental Impact and Land-Use Considerations:

The site selection process should take into account the potential environmental impact of the substation on the surrounding area. Factors such as noise levels, electromagnetic fields, visual impact, and land disturbance should be evaluated. Local regulations and land-use restrictions should be considered to ensure compliance with zoning and environmental requirements.

Land Availability and Size:

- Sufficient land area should be available to accommodate the substation layout, including equipment, clearances, safety zones, and future expansion provisions.
- The site should provide adequate space for fencing, access roads, parking, drainage, and landscaping.

Security:

- The site should be chosen with security considerations in mind. It should be located in a secure area with limited public access to minimize the risk of unauthorized entry and tampering.
- Proximity to security services, such as police and emergency responders, should be considered.

Geotechnical Considerations:

- The site's soil conditions should be evaluated to ensure stable foundation support for the substation structures and equipment.
- Geological assessments should be conducted to assess the potential for seismic activity, landslides, or other geotechnical hazards that may impact the substation's stability and safety.

Environmental Factors:

- The site should be evaluated for potential environmental hazards such as flood zones, high wind areas, or areas prone to extreme weather conditions. Mitigation measures should be considered to address these risks.
- Adequate drainage systems should be in place to prevent waterlogging and ensure proper disposal of stormwater.

Public Perception and Aesthetics:

- The substation should be sited in a location that minimizes visual impact and maintains the aesthetics of the surrounding area.
- Consideration should be given to public perception and community acceptance to minimize potential conflicts and objections.

Grid Connectivity and Network Integration:

- The site should have proximity to existing transmission lines or the ability to establish a connection to the power grid without significant transmission line extensions.
- It should facilitate the integration of the substation into the existing electrical network, ensuring efficient power flow and system stability.

Future Expansion:

- The selected site should allow for future expansion and accommodate the anticipated growth in demand and the addition of new equipment.
- Sufficient space and access to transmission corridors should be available for potential future upgrades and modifications.

Conclusion:

The site selection criteria for substations encompass a range of factors, including proximity to power sources and load centers, accessibility, environmental impact, land availability, security, geotechnical considerations, environmental factors, public perception, grid connectivity, and future expansion provisions. By carefully evaluating and addressing these criteria, the chosen site can support the efficient and reliable operation of the substation while considering environmental, social, and technical aspects. A well-selected site ensures the successful implementation of the substation, enhances system performance, and minimizes potential risks and disruptions in the power grid.

Design Considerations: Layout, Clearances, Grounding, and Safety

The design of an electrical substation involves several critical considerations to ensure efficient operation, reliability, and safety. This section provides an elaborate exploration of key design considerations, including layout, clearances, grounding, and safety measures.

Layout Considerations:

The layout of a substation involves the arrangement and positioning of equipment, structures, and other components. Important factors to consider include:

1. **Equipment Placement:** Proper placement of major equipment, such as transformers, circuit breakers, switchgear, and control panels, is crucial for efficient operation, maintenance access, and safety. Considerations should be given to factors such as clearances, ventilation, and cable routing.
2. **Segregation of Voltage Levels:** Different voltage levels should be physically and electrically segregated within the substation to prevent accidental contact and ensure safety. This segregation helps to minimize the risk of electrical shock and improves system reliability.
3. **Space Allocation:** Sufficient space should be allocated for clearances, safety zones, and future expansion. Clearances should adhere to applicable electrical codes and standards to prevent overheating, ensure proper operation, and facilitate maintenance activities.
4. **Cable Management:** Proper planning for cable routing, cable trenches, conduits, and cable trays is essential to ensure organized and safe cable management. Adequate space should be allocated for cable bends, support structures, and access points to facilitate maintenance and reduce the risk of cable damage.

Clearances:

Clearances refer to the minimum distances required between different equipment, conductors, and structures to prevent electrical breakdown and ensure safety. Considerations for clearances include:

1. **Electrical Clearances:** Clearances should be maintained between conductors, between conductors and grounded surfaces, and between conductors and surrounding

structures. These clearances help to prevent flashovers, electrical faults, and accidental contact.

2. **Safety Clearances:** Safety clearances are required for personnel access, equipment maintenance, and safe operation. They include adequate working space, walkways, access platforms, and sufficient clearance around live parts to protect personnel from accidental contact and ensure safe working conditions.

3. **Minimum Approach Distances:** Minimum approach distances should be followed to protect personnel from electrical hazards. These distances define the closest distance that personnel can approach live parts based on the voltage level and type of equipment.

4. **Code Compliance:** Clearances should adhere to the requirements specified in applicable electrical codes, regulations, and industry standards. National and local electrical codes provide specific guidelines to ensure adequate clearances and promote safe practices.

Grounding:

Grounding systems are essential for the safe and reliable operation of substations. Considerations for grounding include:

1. **Equipment Grounding:** All metallic equipment, enclosures, and structures should be effectively grounded to prevent electrical shocks and ensure proper operation. Equipment grounding provides a low-impedance path for fault currents, facilitating the operation of protective devices and reducing the risk of damage to equipment and personnel.

2. **Grounding Electrodes:** Proper grounding electrodes, such as ground rods, grounding grids, and grounding mats, should be installed to ensure adequate grounding resistance and effective dissipation of fault currents. Soil resistivity and corrosion resistance of grounding components should be considered during design.

3. **Grounding of Non-Current-Carrying Metallic Parts:** Non-current-carrying metallic parts, such as fences, frames, and cable shields, should be bonded and grounded to prevent hazardous voltage differences and ensure equipotential grounding.

4. **Lightning Protection Grounding:** Lightning protection systems should include proper grounding to safely dissipate lightning-induced surges. Grounding electrodes, conductors, and surge protection devices should be installed in accordance with recognized lightning protection standards.

Safety Measures:

Safety measures are crucial to protect personnel, equipment, and the environment. Considerations for safety include:

1. **Signage and Labeling:** Proper signage, warning labels, and equipment markings should be installed to provide clear identification of hazards, operating instructions, and emergency procedures. This ensures that personnel can easily identify and understand potential risks.

2. **Access Control:** Access to substations should be controlled and restricted to authorized personnel only. Fencing, gates, and security systems should be in place to prevent unauthorized access and enhance safety and security.

3. **Fire Protection:** Fire detection and suppression systems, such as fire alarms, extinguishers, and sprinkler systems, should be installed to mitigate fire risks. Proper spacing, ventilation, and fire-resistant construction materials should be considered to contain and prevent the spread of fires.

4. **Personal Protective Equipment (PPE):** The use of appropriate PPE, such as insulated gloves, safety glasses, protective clothing, and footwear, should be promoted and

enforced to ensure personnel safety during construction, maintenance, and operation.

5. **Training and Safety Procedures:** Personnel should receive appropriate training on substation safety, including lockout/tagout procedures, emergency response, and first aid. Safety protocols and procedures should be established and regularly reviewed to maintain a safe working environment.

Conclusion:

Design considerations for substations, including layout, clearances, grounding, and safety measures, are crucial for the efficient and safe operation of electrical systems. Proper layout design ensures optimal equipment placement, cable management, and future expansion provisions. Adequate clearances prevent electrical breakdowns, ensure safe operation and maintenance access, and protect personnel. Effective grounding systems protect against electrical shocks, enhance equipment performance, and dissipate fault currents. Implementing comprehensive safety measures, including signage, access control, fire protection, and personnel training, safeguards personnel, equipment, and the environment. By incorporating these design considerations, substations can operate reliably, reduce downtime, and maintain a safe working environment for personnel.

Environmental and Aesthetic Aspects of Substation Design

In addition to functional considerations, the design of electrical substations should also take into account environmental and aesthetic aspects. Substations are often located in urban or residential areas, and their visual impact and compatibility with the surrounding environment are important factors to consider. This section elaborates on the environmental and aesthetic aspects of substation design.

Environmental Impact:

Substation design should aim to minimize the environmental impact and promote sustainability. Key considerations include:

1. **Visual Impact:** Substations should be designed in a way that minimizes their visual impact on the surrounding landscape. The architectural design and layout should blend harmoniously with the natural or built environment, reducing visual disturbances for nearby residents or scenic areas.
2. **Noise Reduction:** Substations can generate noise, particularly from cooling systems, transformers, and ventilation equipment. Design measures should be implemented to reduce noise emissions, such as selecting quieter equipment, incorporating noise barriers, and utilizing sound-absorbing materials in enclosures.
3. **Vegetation and Landscaping:** Integrating vegetation and landscaping can help soften the appearance of substations and provide visual screening. Trees, shrubs, and greenery can be strategically placed to create buffers and improve the overall aesthetic appeal while providing environmental benefits, such as noise absorption and air pollution mitigation.
4. **Stormwater Management:** Substations should include appropriate stormwater management systems to minimize runoff and prevent soil erosion. Strategies such as retention ponds, permeable surfaces, and sedimentation basins can be employed to control stormwater and promote sustainable water management.

Aesthetic Considerations:

Aesthetics play an important role in substation design, particularly when substations are located in urban or residential areas. Key considerations include:

1. **Architectural Design:** The architectural design of substation buildings and structures should reflect the surrounding architectural styles and blend seamlessly with the visual character of the area. Attention should be given to the selection of materials, color schemes, and building finishes to create an aesthetically pleasing appearance.
2. **Building Enclosures:** Substation buildings can be designed with aesthetically pleasing enclosures to minimize their visual impact. The use of architectural features, such as cladding, decorative panels, or landscaping elements, can help integrate the substation into the surrounding environment.
3. **Fencing and Security:** While security measures are necessary, the design of fencing and security features should be aesthetically pleasing and in harmony with the overall design concept. The selection of materials, finishes, and colors should consider both security needs and visual appeal.
4. **Lighting Design:** Thoughtful lighting design can enhance the visual appearance of substations, particularly during nighttime. The use of strategically placed lighting fixtures, accent lighting, or architectural illumination can highlight design elements and create a visually appealing ambiance.
5. **Art and Public Installations:** Incorporating public art or installations within or around substations can enhance their aesthetic value and promote community engagement. These artistic elements can help transform substations into landmarks and contribute to the cultural and artistic identity of the surrounding area.

6. **Community Engagement:** Engaging with the local community during the design process can help address concerns and gather input. Public consultations, open houses, or design charrettes can provide opportunities for community members to share their perspectives and contribute to the design considerations.

Conclusion:

Considering the environmental and aesthetic aspects of substation design is crucial for creating substations that are visually appealing, compatible with the surrounding environment, and accepted by the local community. By minimizing the environmental impact, incorporating landscaping and vegetation, reducing noise emissions, and integrating pleasing architectural design, substations can seamlessly coexist with their surroundings. Engaging in community consultations and considering public art installations further fosters a sense of ownership and acceptance. Striking a balance between functionality, environmental sustainability, and aesthetics enhances the overall value and acceptance of substations, contributing to a harmonious integration within the urban or residential landscape.

Substation Automation and Control Systems

Substation automation and control systems play a crucial role in the efficient and reliable operation of electrical substations. These systems leverage advanced technologies to monitor, control, and protect the substation equipment and power grid. This section provides an elaborate exploration of substation automation and control systems, including their components, functions, and benefits.

Components of Substation Automation and Control Systems:

1. Supervisory Control and Data Acquisition (SCADA) System: The SCADA system serves as the central hub for monitoring and controlling the substation. It collects data from various sensors and devices, provides real-time visualization of the substation, and enables operators to remotely monitor and control the substation equipment.

2. Remote Terminal Units (RTUs): RTUs are intelligent devices located at different points within the substation. They gather data from sensors, meters, and protection devices and transmit this information to the SCADA system. RTUs also receive control commands from the SCADA system and execute them to operate and control equipment in the substation.

3. Intelligent Electronic Devices (IEDs): IEDs are devices that perform specific functions, such as protection, control, and measurement, within the substation. They acquire data, analyze it, and take appropriate actions based on predefined algorithms. Examples of IEDs include protective relays, digital meters, fault recorders, and voltage regulators.

4. Communication Infrastructure: A reliable and robust communication infrastructure is essential for exchanging data between various components of the automation system. This infrastructure includes communication protocols, networking equipment, and interfaces that allow seamless communication between devices and systems.

Functions of Substation Automation and Control Systems:

1. **Monitoring and Visualization:** Automation systems continuously monitor various parameters, such as voltage, current, temperature, and status of equipment within the substation. They provide real-time visualization of these

parameters through graphical interfaces, allowing operators to assess the substation's health and performance.

2. **Control and Operation:** Automation systems enable remote control and operation of substation equipment. Operators can remotely issue commands to open or close circuit breakers, regulate voltage levels, switch capacitors or reactors, and control other devices. This improves operational efficiency and reduces the need for manual intervention in the field.

3. **Fault Detection and Protection:** Automation systems incorporate protective relays and fault detection algorithms to identify abnormal conditions, such as overcurrent, short circuits, or voltage sags. When a fault is detected, the system initiates protection actions, such as tripping circuit breakers or isolating faulty sections, to minimize damage and maintain the integrity of the power grid.

4. **Data Acquisition and Analysis:** Automation systems collect and analyze data from various sensors and devices within the substation. This data is used for trend analysis, asset management, predictive maintenance, and system optimization. Advanced analytics techniques, such as machine learning algorithms, can be employed to identify patterns, predict equipment failures, and optimize substation performance.

5. **Alarm and Event Management:** Automation systems generate alarms and notifications in response to abnormal conditions, equipment failures, or system events. Operators are alerted to these alarms, allowing them to quickly respond and take appropriate actions to address the issues.

Benefits of Substation Automation and Control Systems:

1. **Enhanced System Reliability:** Automation systems improve the reliability of substations by enabling real-time monitoring, fault detection, and rapid response to abnormal

conditions. They help prevent or mitigate equipment failures, reduce downtime, and enhance the overall reliability of the power grid.

2. **Improved Operational Efficiency:** Automation systems streamline operational processes by automating routine tasks, enabling remote control, and facilitating centralized monitoring. This reduces the need for manual interventions, enhances system efficiency, and optimizes resource allocation.

3. **Faster Fault Detection and Response:** Automation systems enable faster fault detection and quicker response times. They can detect faults within milliseconds, allowing protective devices to isolate faulty sections and restore the system to normal operation promptly. This reduces the impact of faults, improves system stability, and minimizes disruptions in the power grid.

4. **Enhanced Maintenance and Asset Management:** Automation systems facilitate condition-based and predictive maintenance by collecting and analyzing real-time data from substation equipment. This enables maintenance activities to be planned proactively, optimizes equipment life cycles, and reduces maintenance costs.

5. **Improved Safety and Personnel Protection:** Automation systems enhance safety by reducing the need for manual interventions in hazardous conditions. Operators can remotely control equipment, minimizing the risk of electrical accidents. Fault detection algorithms and protective relays also contribute to enhanced personnel protection by quickly isolating faulty sections.

6. **Data-driven Decision Making:** Automation systems provide operators and engineers with accurate and comprehensive data for analysis and decision-making. This data-driven approach enables informed decision-making, facilitates

system optimization, and supports long-term planning and investment strategies.

Conclusion:

Substation automation and control systems play a vital role in optimizing the operation, reliability, and safety of electrical substations. These systems enable real-time monitoring, control, and protection, improving system reliability, reducing downtime, and enhancing operational efficiency. By leveraging advanced technologies, such as SCADA systems, RTUs, IEDs, and robust communication infrastructure, substations can benefit from enhanced fault detection, improved maintenance strategies, and optimized asset management. Substation automation systems empower operators and engineers with real-time data and analytics, supporting data-driven decision-making and contributing to the overall efficiency and performance of the power grid.

CHAPTER 5: POWER SYSTEM PROTECTION AND CONTROL

Chapter 5 focuses on power system protection and control, which is a critical aspect of electrical substations. The protection and control systems ensure the safe and reliable operation of the power system by detecting and isolating faults, coordinating protective devices, and maintaining stable system conditions. This chapter provides an elaborate exploration of power system protection and control, including its principles, components, and functions.

Chapter 5 emphasizes the significance of power system protection and control in maintaining the stability, reliability, and safety of electrical substations. Understanding protection principles, implementing appropriate protective relays, ensuring proper coordination among protective devices, and integrating control and communication systems are essential for effective fault detection, isolation, and system control. The chapter also highlights the growing importance of cybersecurity in protecting substation automation and control systems from cyber threats. By incorporating robust protection and control systems, substations can operate efficiently, prevent disruptions, and safeguard personnel, equipment, and the power grid as a whole.

Importance of protection systems in electrical substations

The importance of protection systems in electrical substations cannot be overstated. These systems are designed to detect and mitigate faults and abnormal conditions in the power system, ensuring the safe and reliable operation of the substation and the overall power grid. This section provides an elaborate exploration of the importance of protection systems in electrical substations.

Equipment and Personnel Safety:

Protection systems are crucial for ensuring the safety of substation equipment and personnel. They detect and isolate faults, such as short circuits, overcurrent, and ground faults, preventing damage to equipment and minimizing the risk of electrical shocks or injuries to personnel. By rapidly identifying and isolating faults, protection systems contribute to a safer working environment for maintenance personnel and substation operators.

System Stability and Reliability:

Protection systems play a vital role in maintaining the stability and reliability of the power system. They monitor critical parameters, such as voltage, current, and frequency, and initiate protective actions when abnormal conditions are detected. By isolating faulty sections and preventing the spread of faults, protection systems help maintain the integrity of the power grid, minimizing disruptions and reducing downtime. This ensures the reliable supply of electricity to consumers and enhances the overall stability of the electrical network.

Equipment and Asset Protection:

Protection systems help protect valuable substation equipment and assets from damage. Faults can lead to excessive currents, overheating, or mechanical stress on equipment, potentially causing permanent damage or even catastrophic failures. By detecting faults and initiating protective actions, protection systems prevent equipment damage and prolong the lifespan of substation assets. This helps avoid costly equipment replacements, reduces maintenance costs, and ensures the efficient utilization of substation resources.

Fault Detection and Localization:

Protection systems are designed to quickly detect and locate faults in the power system. By accurately identifying the location of a fault, protection systems enable prompt responses and targeted interventions. This minimizes the time required for fault isolation, restoration, and troubleshooting, leading to faster system recovery and reduced downtime. The ability to rapidly detect and locate faults is essential for maintaining the overall reliability and availability of the power grid.

Selectivity and Coordination:

Protection systems ensure selectivity and coordination among protective devices. Selectivity refers to the ability to isolate only the faulty section during a fault, while leaving the rest of the power system operational. By coordinating the operation of protective relays, circuit breakers, and other protective devices, protection systems prevent unnecessary tripping of healthy sections and minimize the impact of faults on the power system. This coordination enhances system reliability, reduces disturbances, and improves overall operational efficiency.

Data Acquisition and Analysis:

Protection systems collect and analyze real-time data from various sensors and devices within the substation. This data provides valuable insights into the performance of the power system, enabling engineers and operators to identify trends, diagnose potential issues, and optimize system operation. By leveraging data-driven analytics, protection systems support predictive maintenance strategies, asset management, and system optimization, leading to improved reliability and cost-effective operation.

Compliance with Standards and Regulations:

Protection systems ensure compliance with national and international standards and regulations related to power system protection. These standards outline the minimum requirements for protection coordination, fault detection, and equipment performance. By adhering to these standards, protection systems help ensure that substations meet safety and reliability criteria, enabling regulatory compliance and maintaining the integrity of the power system.

Conclusion:

Protection systems are a fundamental component of electrical substations, providing critical functions such as fault detection, isolation, and system coordination. The importance of protection systems lies in their ability to safeguard equipment, protect personnel, maintain system stability, and ensure the reliable operation of the power grid. By promptly detecting and mitigating faults, protection systems contribute to a safe, reliable, and efficient power supply, reducing downtime, minimizing equipment damage, and enhancing the overall performance of electrical substations.

Types of Relays and Their Applications: Overcurrent, Differential, Distance, and More

Relays are essential components of protection systems in electrical substations. They play a crucial role in detecting faults, abnormal conditions, and providing selective tripping or isolation of faulty sections. This section provides an elaborate exploration of various types of relays commonly used in substations, along with their applications and benefits.

Overcurrent Relays:

Overcurrent relays are among the most common types of relays used in power system protection. They operate based on the magnitude of current flowing through a circuit and are designed to detect and respond to excessive current levels. Applications of overcurrent relays include:

1. Overload Protection: Overcurrent relays protect equipment from thermal damage caused by prolonged overcurrent conditions, such as overloaded transformers, cables, or motors.
2. Short Circuit Protection: Overcurrent relays detect short circuit faults and initiate quick tripping of circuit breakers to isolate faulty sections, preventing further damage to equipment and the power system.
3. Ground Fault Protection: Overcurrent relays with ground fault elements are used to detect ground faults, which occur when a fault path is created between a conductor and the ground. Ground fault protection is crucial for protecting personnel and equipment from electrical shocks.

Differential Relays:

Differential relays operate based on the comparison of current or voltage between two or more points in an electrical circuit. They are primarily used for protecting electrical equipment, such as transformers and generators, against internal faults. Applications of differential relays include:

1. **Transformer Protection:** Differential relays provide primary and backup protection for transformers. They compare the current entering and leaving the transformer windings to detect internal faults, such as winding shorts or insulation failures.
2. **Generator Protection:** Differential relays safeguard generators by comparing the currents at the generator terminals and neutral point. They are sensitive to internal faults within the generator, such as stator or rotor winding faults.
3. **Motor Protection:** Differential relays are also used for protecting large motors. They compare the currents entering and leaving the motor to detect internal faults or abnormal operating conditions.

Distance Relays:

Distance relays measure the impedance or distance between the relay location and the fault point in the power system. They are widely used for transmission line protection and are based on the principle of fault impedance measurement. Applications of distance relays include:

1. **Transmission Line Protection:** Distance relays are used to detect and localize faults on transmission lines. By measuring the impedance between the relay location and the fault point, they can accurately determine the fault location, allowing for targeted fault isolation.
2. **Pilot Protection:** Distance relays are also used in pilot protection schemes, where relays at each end of a

transmission line communicate to provide backup protection in case of primary protection failure.

Differential and Percentage Differential Relays:

Differential and percentage differential relays are variants of differential relays that provide more precise and selective protection for specific applications. They are commonly used for transformer protection, particularly in high-voltage substations. Applications of differential and percentage differential relays include:

➢ **Transformer Differential Protection:** Differential and percentage differential relays provide highly sensitive and selective protection for transformers. They detect even small internal faults by comparing the currents entering and leaving the transformer windings, allowing for rapid isolation of faulty sections.

Directional Relays:

Directional relays are designed to detect the direction of current flow in a power system. They are used to provide selective tripping or protection coordination in systems with multiple sources or interconnected grids. Applications of directional relays include:

1. **Network Protection:** Directional relays are employed in interconnected power systems to ensure selective tripping of circuit breakers and coordination with other protective devices, preventing faults from propagating across the network.

2. **Generator Protection:** Directional relays can be used to detect reverse power flow, which occurs when a generator is inadvertently connected to the power system and starts

consuming power instead of supplying it. Reverse power flow protection helps prevent generator damage and system instability.

Frequency Relays:

Frequency relays monitor the frequency of the power system and initiate protective actions if the frequency deviates from the normal operating range. Applications of frequency relays include:

1. **Load Shedding:** Frequency relays are used to detect underfrequency conditions in the power system, which may occur due to sudden load imbalances or generation losses. They initiate load shedding schemes to maintain system stability by shedding non-critical loads.
2. **Overfrequency Protection:** Frequency relays can also detect overfrequency conditions caused by excess generation or loss of load. They initiate protective actions to stabilize the system by reducing generation or shedding excess load.

Voltage Relays:

Voltage relays monitor voltage levels in the power system and initiate protective actions if the voltage exceeds or falls below specified thresholds. Applications of voltage relays include:

1. **Undervoltage Protection:** Voltage relays are used to detect undervoltage conditions, which may occur due to faults, load shedding, or voltage regulator malfunctions. They initiate protective actions, such as tripping circuit breakers or activating backup power sources.
2. **Overvoltage Protection:** Voltage relays can also detect overvoltage conditions caused by voltage surges or voltage regulator malfunctions. They initiate protective actions to prevent equipment damage and maintain system stability.

Conclusion:

The different types of relays in electrical substations serve specific purposes and play a vital role in protecting equipment, isolating faults, and maintaining the reliability and stability of the power system. From overcurrent relays for general fault detection to differential, distance, directional, frequency, and voltage relays for more specific applications, each type of relay contributes to the overall protection and coordination of the substation. By implementing appropriate relays and protective schemes, substations can detect and respond to faults swiftly, minimize disruptions, and ensure the safe and reliable operation of the power grid.

Protection Schemes for Transformers, Generators, Transmission Lines, and Other Elements

Protection schemes are essential in electrical substations to safeguard various elements of the power system, such as transformers, generators, transmission lines, and other equipment. These schemes employ a combination of relays, sensors, and protective devices to detect faults, isolate faulty sections, and ensure the reliable operation of the power grid. This section provides an elaborate exploration of the protection schemes commonly used for transformers, generators, transmission lines, and other elements in substations.

Transformer Protection Schemes:

Transformers are critical components in electrical substations, and protection schemes are employed to ensure their safe and reliable operation. Some common protection schemes for transformers include:

1. **Differential Protection:** Differential protection is widely used for transformer protection. Current transformers (CTs) are connected at the primary and secondary windings of the

transformer, and the differential relay compares the currents. Any imbalance in the currents indicates an internal fault within the transformer, leading to the tripping of the associated circuit breaker.

2. **Buchholz Relay Protection:** Buchholz relays are gas and oil-actuated relays placed in the oil-filled conservator tank of a transformer. They detect faults such as internal short circuits or overheating by sensing gas or oil flow changes due to fault-related gas generation or oil movement. Buchholz relays initiate an alarm or trip the transformer if a fault is detected.

3. **Overcurrent Protection:** Overcurrent relays provide protection against excessive currents in the transformer windings caused by faults or overloads. These relays detect the current magnitude and trip the associated circuit breaker if the current exceeds a predetermined threshold.

4. **Restricted Earth Fault (REF) Protection:** REF protection is used to detect earth faults within the transformer windings. It relies on current transformers placed on the transformer neutral and winding connection points. If a fault occurs within the windings, a current imbalance is detected, and the REF relay initiates a trip signal to isolate the faulty section.

Generator Protection Schemes:

Generators are vital components of power systems, and protection schemes are implemented to ensure their safe and reliable operation. Some common protection schemes for generators include:

1. **Differential Protection:** Differential protection schemes are commonly used for generator stator and rotor windings. Current transformers are connected at the generator

terminals and the neutral point, and the differential relay compares the currents. Any imbalance in the currents indicates an internal fault within the generator, leading to the tripping of the associated circuit breaker.

2. **Loss of Field (LOF) Protection:** LOF protection is used to detect a loss of excitation or field current in a generator. It monitors the excitation system and initiates a trip signal if the field current falls below a specified threshold, ensuring that the generator is disconnected from the grid to prevent damage.

3. **Overvoltage Protection:** Overvoltage relays monitor the generator's output voltage and initiate protective actions if the voltage exceeds a predetermined threshold. Overvoltage protection prevents excessive voltage levels that can lead to insulation failure or damage to connected equipment.

4. **Overfrequency and Underfrequency Protection:** Frequency relays monitor the generator's output frequency and initiate protective actions if the frequency exceeds or falls below predetermined thresholds. These relays help maintain system stability and protect the generator and connected devices from damage caused by abnormal frequency conditions.

Transmission Line Protection Schemes:

Transmission lines are crucial components of the power system, and protection schemes are implemented to ensure their reliable operation and prevent catastrophic failures. Some common protection schemes for transmission lines include:

1. **Distance Protection:** Distance relays are widely used for transmission line protection. They measure the impedance or distance to the fault location and initiate tripping of the circuit breaker if a fault occurs within a specific distance

from the relay. Distance protection ensures selective fault isolation and minimizes disruption to the power system.

2. **Overcurrent Protection:** Overcurrent relays provide backup protection for transmission lines. They detect excessive currents caused by faults or overloads and initiate tripping of the associated circuit breaker to isolate the faulty section.

3. **Pilot Protection:** Pilot protection schemes use communication-based relaying systems to detect faults on transmission lines. These schemes involve communication between relays at each end of the line, allowing for precise fault detection and selective tripping.

4. **Directional Overcurrent Protection:** Directional overcurrent relays are used to protect transmission lines from faults originating from specific directions. They provide directional sensing and initiate tripping only if a fault occurs from the specified direction, preventing unnecessary tripping for faults outside the protection zone.

Other Element Protection Schemes:

There are several other protection schemes employed in substations to protect various elements of the power system. Some examples include:

1. **Motor Protection:** Motor protection schemes involve a combination of overcurrent, differential, and thermal protection techniques to safeguard motors from faults and abnormal operating conditions.

2. **Busbar Protection:** Busbar protection schemes use differential protection to detect and isolate faults occurring within the busbar system. These schemes ensure the reliable operation and protection of busbar connections.

3. **Capacitor Bank Protection:** Capacitor bank protection schemes protect capacitor banks from faults and abnormal

operating conditions. These schemes involve overcurrent, overvoltage, and overtemperature protection techniques.

4. **Breaker Failure Protection:** Breaker failure protection schemes monitor the operation of circuit breakers and initiate backup protection actions if the primary breaker fails to trip during a fault. This ensures fault isolation and minimizes disruptions in the power system.

Conclusion:

Protection schemes are crucial for the safe and reliable operation of various elements in electrical substations. These schemes employ a combination of relays, sensors, and protective devices to detect faults, isolate faulty sections, and maintain the stability of the power grid. By implementing appropriate protection schemes for transformers, generators, transmission lines, and other elements, substations can ensure the efficient and secure operation of the power system, minimize downtime, and protect personnel and equipment from damage.

Substation Monitoring, SCADA Systems, and Remote Control

In modern electrical substations, monitoring, supervisory control, and data acquisition (SCADA) systems are widely employed to enable efficient operation, enhance system reliability, and facilitate remote control of substation equipment. This section provides an elaborate exploration of substation monitoring, SCADA systems, and remote control, including their functions, components, and benefits.

Substation Monitoring:

Substation monitoring involves the continuous monitoring of various parameters within the substation to assess the health, performance, and condition of the equipment and the power system. Some key aspects of substation monitoring include:

1. **Real-time Data Acquisition:** Monitoring systems collect real-time data from sensors, meters, and devices within the substation. This data includes voltage levels, current flows, power factor, temperature, and other critical parameters.
2. **Equipment Condition Monitoring:** Monitoring systems continuously assess the condition of substation equipment, such as transformers, circuit breakers, and switchgear. They monitor parameters like temperature, oil level, gas content, vibration, and insulation condition, allowing for predictive maintenance and early fault detection.
3. **Alarm and Event Management:** Monitoring systems generate alarms and notifications in response to abnormal conditions, equipment failures, or system events. These alarms alert operators to take necessary actions promptly.

4. **Trend Analysis and Reporting:** Monitoring systems analyze historical data and generate reports on the performance and trends of the substation. These reports provide insights into the behavior of the power system, identify potential issues, and support decision-making for maintenance and system optimization.

SCADA Systems:

SCADA systems are central components of substation automation that integrate data acquisition, control, and monitoring functions. These systems enable centralized supervisory control and real-time visualization of the substation. Key aspects of SCADA systems include:

1. **Data Acquisition and Visualization:** SCADA systems gather data from sensors, meters, and devices within the substation and provide real-time visualization of the substation layout and parameters. This visualization

includes graphical representations of equipment, status indications, and trends of critical parameters.

2. **Remote Control and Operation:** SCADA systems enable remote control and operation of substation equipment. Operators can issue commands to open or close circuit breakers, regulate voltage levels, switch capacitors or reactors, and control other devices from a central control room.

3. **Alarms and Event Logging:** SCADA systems generate alarms and event logs to alert operators of abnormal conditions, equipment failures, or system events. These features provide a comprehensive overview of the substation's operational status and facilitate quick response and troubleshooting.

4. **Data Storage and Historical Analysis:** SCADA systems store historical data for further analysis and reporting. This data can be used to identify trends, diagnose faults, optimize system operation, and support regulatory compliance and reporting requirements.

5. **Integration with Other Systems:** SCADA systems can integrate with other systems, such as protection relays, energy management systems, and enterprise resource planning systems. This integration allows for seamless data exchange, coordination, and optimization of substation operation within the broader power system.

Remote Control and Operation:

Remote control and operation enable operators to control and monitor substation equipment from a centralized location, typically the control room. Key aspects of remote control and operation include:

1. **Control Room Operations:** Operators in the control room can remotely control various substation equipment, such as circuit breakers, transformers, and capacitor banks. They

can execute control actions based on real-time data and system requirements.

2. **Enhanced Safety:** Remote control and operation reduce the need for manual interventions in hazardous areas of the substation, improving personnel safety. Operators can remotely perform switching operations and equipment control without physical presence in the field.

3. **Efficiency and Optimization:** Remote control and operation streamline operational processes, minimize response times, and enhance overall operational efficiency. Operators can remotely monitor and control multiple substations, reducing travel time and resource requirements.

4. **Fault Management and Restoration:** Remote control and operation facilitate quick fault detection, isolation, and restoration of the power system. Operators can remotely initiate switching operations, isolate faulty sections, and restore service to minimize downtime.

5. **Flexibility and Scalability:** Remote control and operation enable efficient management of multiple substations and allow for easy scalability as the power system expands. Operators can remotely monitor and control substations located in different geographical areas, providing flexibility in system management.

Benefits of Substation Monitoring, SCADA Systems, and Remote Control:

The implementation of substation monitoring, SCADA systems, and remote control offers several benefits, including:

1. **Enhanced System Reliability:** Continuous monitoring and real-time data acquisition enable early fault detection, minimizing downtime and improving overall system reliability.

2. **Improved Operational Efficiency:** SCADA systems and remote control streamline operational processes, reduce

response times, and optimize resource allocation, leading to improved efficiency in substation operation.

3. **Remote Troubleshooting and Maintenance:** Remote control capabilities facilitate remote troubleshooting and maintenance, reducing the need for physical presence in the field and optimizing maintenance activities.

4. **Enhanced Safety and Personnel Protection:** Remote control and operation minimize the need for manual interventions in hazardous areas, improving personnel safety and reducing the risk of electrical accidents.

5. **Data-driven Decision Making:** Substation monitoring and SCADA systems provide accurate and comprehensive data for analysis and decision-making. This data-driven approach supports predictive maintenance, system optimization, and long-term planning.

6. **Improved Grid Management:** Substation automation and remote control enable coordinated and optimized operation of multiple substations, contributing to efficient grid management and stability.

Conclusion:

Substation monitoring, SCADA systems, and remote control have become integral components of modern electrical substations. These technologies enable real-time data acquisition, centralized control, and remote operation of substation equipment. By providing accurate information, enhancing system reliability, improving operational efficiency, and ensuring personnel safety, these systems contribute to the effective and reliable operation of the power grid. The integration of monitoring, SCADA systems, and remote control enables efficient grid management, predictive maintenance, and data-driven decision-making for optimal substation performance.

CHAPTER 6: SUBSTATION MAINTENANCE AND ASSET MANAGEMENT

Chapter 6 focuses on the important aspects of substation maintenance and asset management. Effective maintenance practices and asset management strategies are crucial for ensuring the reliable and efficient operation of electrical substations. This chapter provides an elaborate exploration of substation maintenance and asset management, including their significance, key activities, and benefits.

Chapter 6 highlights the importance of substation maintenance and asset management for ensuring the reliable and efficient operation of electrical substations. By implementing effective maintenance strategies, conducting routine inspections, utilizing advanced condition monitoring techniques, and following industry best practices, substations can optimize

73

equipment performance, extend asset lifespan, and minimize downtime. Additionally, adopting asset management principles and utilizing the right tools and technologies contribute to data-driven decision-making, improved asset utilization, and overall substation reliability. By prioritizing maintenance, adhering to regulatory requirements, and fostering a culture of safety, substations can maximize equipment reliability, enhance system availability, and ensure a safe working environment for maintenance personnel.

Importance of Regular Maintenance and Inspections

Regular maintenance and inspections are of utmost importance in electrical substations to ensure the reliable and safe operation of the power system. These activities involve systematic checks, testing, and servicing of substation equipment to identify potential issues, prevent failures, and maintain optimal performance. This section provides an elaborate exploration of the importance of regular maintenance and inspections in electrical substations.

Equipment Reliability:

Regular maintenance and inspections help maintain equipment reliability. Substation equipment, such as transformers, circuit breakers, and switchgear, are subject to various stresses, environmental factors, and operational demands. Without proper maintenance, these components can deteriorate over time, leading to increased failure rates and unexpected outages. By conducting regular inspections and maintenance activities, potential problems can be identified and addressed proactively, minimizing the risk of equipment failure and ensuring the long-term reliability of the power system.

Early Fault Detection:

Regular inspections and maintenance activities facilitate early fault detection. By performing visual inspections, conducting tests, and analyzing equipment performance, potential faults or abnormalities can be identified before they escalate into more significant issues. Early detection allows for timely intervention and corrective measures, reducing the likelihood of catastrophic failures, system disruptions, and costly repairs. It enables maintenance personnel to identify deteriorating components, loose connections, insulation breakdown, or other potential failure points, ensuring prompt repairs or replacements.

System Availability and Downtime Reduction:

Regular maintenance and inspections contribute to improved system availability and reduced downtime. Unplanned outages and equipment failures can disrupt the power supply, affecting consumers, businesses, and critical services. By proactively maintaining and inspecting substation equipment, potential issues can be addressed before they lead to failures or interruptions in service. This proactive approach minimizes downtime, optimizes system availability, and enhances the overall reliability of the power grid.

Safety Enhancement:

Regular inspections and maintenance activities promote a safer working environment for substation personnel. Electrical substations involve high voltages, potential hazards, and complex equipment. Regular maintenance ensures that safety devices, such as grounding systems and protective relays, are functioning properly. It also allows for the identification and mitigation of safety risks, such as deteriorating insulation, loose connections, or faulty protective devices. By addressing safety concerns through regular maintenance and inspections, the risk of accidents, electrical shocks, and injuries to personnel is significantly reduced.

Compliance with Standards and Regulations:

Regular maintenance and inspections ensure compliance with industry standards and regulatory requirements. Electrical substations are subject to various standards and regulations that outline maintenance practices, testing protocols, and safety guidelines. Regular inspections and maintenance activities help verify compliance with these requirements, ensuring that substations meet the necessary safety, performance, and reliability criteria. Compliance with standards and regulations not only ensures the proper functioning of substation equipment but also helps prevent penalties, legal issues, and reputational damage.

Cost Optimization:

Regular maintenance and inspections contribute to cost optimization. While maintenance activities incur expenses, they help identify minor issues before they escalate into major failures. By addressing small problems proactively, the need for extensive repairs or replacements can be minimized. Regular maintenance also prolongs the lifespan of substation equipment, avoiding premature equipment replacements and reducing long-term costs. Additionally, planned maintenance activities can be scheduled during low-demand periods, minimizing the impact on electricity consumers.

Asset Management:

Regular maintenance and inspections are integral to effective asset management. By monitoring and assessing equipment performance, maintenance activities help optimize the utilization of substation assets. They provide critical data for asset management strategies, such as condition assessment, risk analysis, and lifecycle management. By understanding the condition of assets through regular inspections, decisions regarding repairs, replacements, or upgrades can be made based on data-driven assessments, optimizing asset performance and longevity.

Conclusion:

Regular maintenance and inspections are paramount for the reliable and safe operation of electrical substations. They ensure equipment reliability, facilitate early fault detection, enhance system availability, promote a safe working environment, and enable compliance with standards and regulations. Through regular maintenance, substations can optimize costs, extend asset lifespan, and improve overall power system performance. By prioritizing regular maintenance and inspections, substations can prevent costly failures, reduce downtime, and maintain a robust and efficient electrical infrastructure.

Asset Management Strategies for Substations

Asset management strategies are essential for maintaining the reliability, performance, and longevity of assets in electrical substations. These strategies encompass a range of activities aimed at optimizing asset utilization, minimizing risks, and ensuring cost-effective operation. This section provides an elaborate exploration of asset management strategies specifically tailored to substations.

Asset Inventory and Data Management:

Effective asset management begins with a comprehensive asset inventory and data management system. This involves creating an accurate and up-to-date inventory of all substation assets, including transformers, circuit breakers, switchgear, relays, and other components. The inventory should include relevant details such as manufacturer, model, serial numbers, installation dates, and maintenance history. A well-structured data management system ensures easy access to asset information and facilitates efficient tracking, maintenance planning, and decision-making.

Asset Condition Assessment:

Regular and systematic condition assessment of substation assets is critical for effective asset management. Condition assessment involves evaluating the health and performance of assets through inspections, testing, and monitoring. This assessment provides insights into asset condition, identifies potential risks, and supports proactive maintenance strategies. Techniques such as visual inspections, thermography, oil analysis, and electrical testing help assess the condition of assets and determine appropriate maintenance actions.

Risk Analysis and Prioritization:

Risk analysis is a crucial aspect of asset management for substations. It involves evaluating risks associated with asset failures, potential impacts on system reliability, safety hazards, and financial consequences. By conducting risk assessments, assets can be prioritized based on criticality, consequence of failure, and likelihood of occurrence. This allows for targeted maintenance, resource allocation, and risk mitigation strategies to minimize the overall risk exposure of the substation.

Lifecycle Management:

Lifecycle management involves planning for the entire lifecycle of substation assets, from installation to retirement. It includes strategies for asset acquisition, operation, maintenance, and decommissioning. By considering the complete lifecycle, substations can optimize asset utilization, minimize costs, and maximize asset lifespan. Key considerations in lifecycle management include asset design, procurement, installation, maintenance practices, and end-of-life planning.

Maintenance Strategies:

Maintenance strategies play a crucial role in asset management for substations. They include preventive, predictive, and corrective maintenance approaches to ensure optimal asset performance. Preventive maintenance involves routine inspections, servicing, and component replacements based on predetermined schedules. Predictive maintenance utilizes condition monitoring techniques, data analysis, and trend analysis to identify potential failures and optimize maintenance intervals. Corrective maintenance is carried out in response to equipment failures or abnormal conditions. An effective maintenance strategy combines these approaches to balance costs, equipment reliability, and system availability.

Spare Parts Management:

An effective spare parts management strategy is vital for minimizing downtime and ensuring timely repairs. This strategy involves maintaining an inventory of critical spare parts, such as circuit breakers, relays, fuses, and other components. Spare parts should be properly stored, tracked, and regularly reviewed for accuracy and obsolescence. A proactive approach to spare parts management ensures quick replacements, minimizes downtime, and improves overall asset availability.

Performance Monitoring and Reporting:

Performance monitoring and reporting are essential components of asset management strategies. Key performance indicators (KPIs) are established to measure asset performance, reliability, maintenance costs, and system availability. Regular performance monitoring allows for trend analysis, benchmarking, and identification of areas for improvement. Performance reports provide insights into asset performance, maintenance effectiveness, and support data-driven decision-making for optimizing asset management practices.

Technological Integration:

Technological integration plays a significant role in modern asset management strategies for substations. Advanced technologies such as condition monitoring systems, predictive analytics, asset management software, and remote monitoring capabilities enable efficient data collection, analysis, and decision-making. Integration of these technologies enhances asset visibility, facilitates proactive maintenance, optimizes resource allocation, and supports long-term asset planning.

Training and Knowledge Management:

Investing in training and knowledge management initiatives is critical for effective asset management. Training programs enhance the technical skills and knowledge of maintenance personnel, ensuring proper handling, inspection, and maintenance of substation assets. Knowledge management involves capturing, organizing, and sharing best practices, lessons learned, and asset-related information across the organization. This promotes consistent and standardized asset management practices and enables continuous improvement.

Conclusion:

Effective asset management strategies are vital for the reliable, safe, and cost-effective operation of substations. By implementing asset inventory and data management systems, conducting regular condition assessments, performing risk analysis, and employing appropriate maintenance strategies, substations can optimize asset performance, minimize downtime, and mitigate risks. A comprehensive asset management approach ensures the longevity of substation assets, maximizes asset utilization, and supports the overall goals of the power system. By prioritizing asset management, substations can achieve operational excellence, enhance system reliability, and ensure the efficient utilization of resources.

Diagnostic Techniques for Equipment Condition Monitoring

Effective equipment condition monitoring is essential for maintaining the reliability, performance, and longevity of substation equipment. Diagnostic techniques play a crucial role in identifying potential issues, assessing the health of equipment, and enabling proactive maintenance actions. This section provides an elaborate exploration of diagnostic techniques commonly used for equipment condition monitoring in electrical substations.

Visual Inspection:

Visual inspection is a fundamental diagnostic technique used to assess the condition of equipment. It involves visually examining equipment for signs of wear, corrosion, loose connections, physical damage, and other visible abnormalities. Visual inspection can provide valuable information about the overall condition of equipment and identify any immediate issues that require attention. It is a relatively simple and cost-effective technique that can be performed regularly by trained personnel.

Thermography:

Thermography, or infrared thermographic inspection, is a non-contact diagnostic technique used to detect abnormal temperature variations in electrical equipment. It involves using thermal imaging cameras to capture and analyze the heat patterns emitted by equipment. Thermography can identify overheating components, loose connections, faulty insulation, and other thermal anomalies that may indicate potential problems. By detecting these anomalies early, thermography enables the identification of potential faults and the implementation of corrective actions before failures occur.

Electrical Testing:

Electrical testing encompasses a range of diagnostic techniques used to evaluate the electrical characteristics and performance of equipment. Some common electrical testing techniques include:

1. **Insulation Resistance Testing:** Insulation resistance testing measures the resistance of insulation materials to detect insulation degradation or breakdown. It helps identify potential faults, moisture ingress, or aging insulation that may lead to equipment failures or electrical hazards.

2. **Power Factor Testing:** Power factor testing assesses the power factor of equipment, such as transformers and capacitors. Deviations from the expected power factor can indicate insulation issues, internal faults, or deteriorating components.

3. **Partial Discharge Testing:** Partial discharge testing detects and measures partial discharges within electrical equipment. Partial discharge activity can be an early indication of insulation degradation or impending equipment failures.

4. **Dielectric Withstand Testing:** Dielectric withstand testing evaluates the ability of equipment to withstand high voltage stresses without breakdown. It ensures the insulation integrity and overall reliability of equipment.

Oil Analysis:

Oil analysis is commonly used for assessing the condition of oil-filled equipment, such as transformers and circuit breakers. It involves analyzing samples of the insulating oil to evaluate its chemical and physical properties. Oil analysis can detect the presence of contaminants, moisture, degradation by-products, and insulation degradation. By monitoring the oil condition, potential issues can be identified early, and appropriate maintenance actions can be taken to prevent equipment failures or malfunctions.

Vibration Analysis:

Vibration analysis is a diagnostic technique used to assess the mechanical condition of rotating equipment, such as motors and generators. It involves measuring and analyzing the vibration levels and patterns to identify any abnormalities or excessive vibration that may indicate potential bearing wear, misalignment, or other mechanical issues. Vibration analysis enables the detection of early signs of equipment deterioration, allowing for timely corrective actions.

Dissolved Gas Analysis (DGA):

Dissolved Gas Analysis is a technique used specifically for assessing the condition of transformer insulation oil. It involves analyzing the dissolved gases in the oil to detect the presence of certain gas ratios or abnormal gas concentrations. The presence of specific gases, such as hydrogen, methane, ethylene, and acetylene, can indicate various types of faults, such as overheating, arcing, or insulation degradation within the transformer. DGA helps identify potential faults and enables timely maintenance or replacement actions.

Ultrasonic Testing:

Ultrasonic testing involves the use of ultrasonic waves to detect and analyze the presence of high-frequency sound waves emitted by equipment. It is particularly useful for detecting and locating partial discharges, arcing, corona, and other high-frequency phenomena that may occur within electrical equipment. Ultrasonic testing can identify potential insulation issues, loose connections, and other problems that may lead to equipment failures or malfunctions.

Data Analytics and Trend Analysis:

Data analytics and trend analysis utilize advanced algorithms and software to analyze large volumes of data collected from various diagnostic techniques. These techniques help identify patterns, anomalies, and trends in equipment performance, condition, and failure modes. By analyzing historical data and comparing current measurements, potential issues can be detected early, enabling proactive maintenance planning and resource allocation.

Conclusion:

Diagnostic techniques for equipment condition monitoring are critical for identifying potential issues, assessing equipment health, and enabling proactive maintenance actions in electrical substations. Visual inspection, thermography, electrical testing, oil analysis, vibration analysis, dissolved gas analysis, ultrasonic testing, and data analytics all contribute to effective equipment condition monitoring. By employing these diagnostic techniques, substations can detect abnormalities, mitigate risks, prevent equipment failures, and optimize maintenance strategies. These techniques enable substations to proactively address equipment issues, enhance equipment reliability, and ensure the reliable operation of the power system.

Life Cycle Management and Refurbishment of Substations

Life cycle management and refurbishment strategies are crucial for the effective operation, maintenance, and optimization of electrical substations. Substations undergo aging, technological advancements, changing operational requirements, and evolving regulations over their lifespan. Proper life cycle management ensures that substations are maintained, upgraded, and refurbished to meet current and future demands. This section provides an elaborate exploration of life cycle management and refurbishment strategies for substations.

Life Cycle Management:

Life cycle management encompasses a holistic approach to managing substations throughout their entire life cycle, from planning and design to decommissioning. It involves considering various aspects, including asset acquisition, operation, maintenance, refurbishment, and end-of-life considerations. Key components of life cycle management include:

1. **Planning and Design:** Proper planning and design at the early stages of a substation's life cycle lay the foundation for efficient operation, maintenance, and future refurbishment. Factors such as location, layout, capacity, future expansion requirements, and regulatory compliance are considered during the planning and design phase.

2. **Asset Acquisition:** Careful selection and acquisition of substation assets ensure their suitability, reliability, and compatibility with the substation's requirements. Factors such as equipment quality, performance, supplier reputation, and long-term support should be considered during the acquisition process.

3. **Operation and Maintenance:** Regular operation and maintenance activities, as discussed in previous chapters, ensure the optimal performance and reliability of substations. Adhering to maintenance schedules, conducting condition assessments, and implementing preventive and predictive maintenance strategies are essential for effective life cycle management.

4. **Refurbishment and Upgrades:** Refurbishment and upgrades are undertaken to extend the lifespan, improve performance, and meet changing operational requirements of substations. These activities may include equipment replacements, technology upgrades, capacity expansions, and compliance with new regulations. Refurbishment should be based on a cost-benefit analysis considering factors such as asset condition, performance improvements, energy efficiency, and long-term operational costs.

5. **Decommissioning and Disposal:** At the end of a substation's life cycle, proper decommissioning and disposal procedures must be followed. This involves safely disconnecting equipment, managing hazardous materials, and ensuring compliance with environmental regulations. Decommissioning should be planned well in advance, considering the impact on the power system, environmental implications, and cost-effective disposal methods.

Refurbishment Strategies:

Refurbishment strategies are employed to enhance the performance, reliability, and safety of existing substations. They involve targeted upgrades and modifications to address specific issues, improve efficiency, and prolong asset lifespan. Key considerations for refurbishment strategies include:

1. **Equipment Replacement:** Aging or obsolete equipment that poses reliability concerns or hampers substation performance may require replacement. This could involve replacing transformers, circuit breakers, switchgear, protective relays, or other critical components with modern, more efficient, and technologically advanced equipment.

2. **Technology Upgrades:** Technological advancements often provide opportunities for upgrading substation systems. Upgrades may include replacing electromechanical relays with digital relays, implementing advanced automation and control systems, adopting new communication protocols, or integrating intelligent monitoring and diagnostic systems.

3. **Capacity Expansion:** Increasing electricity demand may necessitate capacity expansion in substations. This could involve adding new equipment, expanding the footprint of the substation, or upgrading existing equipment to handle higher voltages or current loads.

4. **Compliance with Regulations:** Refurbishment strategies should consider compliance with evolving safety, environmental, and regulatory standards. This may involve upgrading insulation systems, enhancing grounding systems, improving fire protection measures, or adopting new safety features.

5. **Energy Efficiency Improvements:** Refurbishment can include measures to improve energy efficiency, reduce losses, and optimize power consumption. This may involve implementing energy-efficient transformers, upgrading insulation materials, optimizing cooling systems, or installing renewable energy generation systems within the substation.

6. **Condition-Based Refurbishment:** Condition-based refurbishment focuses on addressing specific equipment issues identified through condition monitoring and

assessment techniques. By targeting the root cause of problems, refurbishment efforts can be directed where they are most needed, optimizing resource utilization and minimizing downtime.

7. **Future-Proofing:** Refurbishment strategies should consider future technological advancements, changing regulatory requirements, and anticipated operational demands. Future-proofing involves implementing flexible designs, scalable solutions, and considering long-term reliability and maintainability of refurbished assets.

Economic Considerations:

Life cycle management and refurbishment decisions must consider economic factors, including costs, benefits, and return on investment. A comprehensive cost-benefit analysis helps evaluate the financial viability of refurbishment projects, considering factors such as equipment lifespan extension, energy savings, operational efficiency improvements, maintenance costs, and potential revenue loss due to downtime during refurbishment.

Risk Assessment:

Risk assessment plays a vital role in life cycle management and refurbishment strategies. It involves identifying and assessing risks associated with aging equipment, technological obsolescence, changing operational demands, safety hazards, and compliance with regulations. Risk assessments help prioritize refurbishment activities, allocate resources effectively, and mitigate potential risks to ensure the continued reliable and safe operation of substations.

Conclusion:

Life cycle management and refurbishment strategies are essential for ensuring the optimal performance, reliability, and longevity of electrical substations. By adopting a comprehensive life cycle approach, substations can address equipment aging, upgrade technologies, meet changing operational requirements, and comply with evolving regulations. Refurbishment strategies should be based on a thorough analysis of equipment condition, operational needs, economic considerations, and risk assessments. Proper life cycle management and targeted refurbishment efforts ensure the efficient utilization of assets, enhance substation performance, and contribute to the reliability and sustainability of the power system.

CHAPTER 7: EMERGING TECHNOLOGIES AND FUTURE TRENDS

Chapter 7 explores the emerging technologies and future trends that are shaping the landscape of electrical substations. As the power industry evolves, new technologies are being developed to enhance the efficiency, reliability, and sustainability of substations. This chapter provides an elaborate exploration of the emerging technologies and future trends that hold promise for substations.

Chapter 7 highlights the emerging technologies and future trends that are shaping the evolution of electrical substations. Digital substations, IoT, big data analytics, grid automation, renewable energy integration, energy storage systems, cybersecurity, electrification, and AI are transforming the way substations operate and interact with the power system. These technologies hold promise for improving substation efficiency, enhancing grid reliability, enabling renewable energy integration, and supporting sustainable power systems. By embracing these emerging technologies and keeping abreast of future trends, substations can adapt to the changing energy landscape and contribute to the development of a smart, resilient, and sustainable power grid.

Integration of Renewable Energy Sources in Substations

The integration of renewable energy sources in substations plays a crucial role in transitioning towards a sustainable and low-carbon energy future. Renewable energy sources, such as solar, wind, and biomass, offer environmentally friendly alternatives to traditional fossil fuel-based power generation. This section provides an elaborate exploration of the integration of renewable energy sources in substations, including the challenges, solutions, and benefits associated with this integration.

Importance of Renewable Energy Integration:

The integration of renewable energy sources in substations is vital for several reasons:

1. Environmental Sustainability: Renewable energy sources produce electricity with minimal greenhouse gas emissions, reducing the carbon footprint and promoting environmental sustainability.
2. Diversification of Energy Sources: Integrating renewable energy sources diversifies the energy mix, reducing dependence on fossil fuels and enhancing energy security.
3. Meeting Renewable Energy Targets: Many countries have set ambitious renewable energy targets, and the integration of renewables in substations is essential for achieving these goals.
4. Decentralized Power Generation: Renewable energy sources often have a distributed and decentralized nature, allowing for local power generation and reducing transmission losses.
5. Economic Benefits: The integration of renewable energy sources can stimulate economic growth, create jobs, and attract investment in the renewable energy sector.

Challenges of Renewable Energy Integration:

The integration of renewable energy sources in substations presents various challenges:

1. **Intermittency and Variability:** Renewable energy sources, such as solar and wind, are intermittent and variable in nature, depending on weather conditions. This requires careful management and coordination to ensure a stable and reliable power supply.
2. **Grid Stability and Power Quality:** Integrating intermittent renewable energy sources can impact grid stability and power quality. Fluctuations in generation must be managed

to maintain grid voltage and frequency within acceptable limits.

3. **Grid Capacity and Reinforcement:** Substation infrastructure may need to be upgraded or reinforced to accommodate the increased capacity and variability associated with renewable energy integration. This may involve the installation of additional transformers, circuit breakers, and other equipment.

4. **Grid Connection and Interconnection:** Connecting renewable energy sources to the grid requires appropriate interconnection infrastructure and compliance with grid codes and regulations.

5. **Forecasting and Predictability:** Accurate forecasting and prediction of renewable energy generation are crucial for grid planning, dispatch, and balancing. Advanced forecasting techniques and data analytics are needed to optimize system operations.

Solutions for Renewable Energy Integration:

Several solutions and strategies are employed to address the challenges of integrating renewable energy sources in substations:

1. **Grid Codes and Standards:** Grid codes specify the technical requirements and guidelines for connecting renewable energy sources to the grid. Compliance with grid codes ensures the safe and reliable operation of the power system.

2. **Energy Management Systems:** Energy management systems provide real-time monitoring, control, and optimization of renewable energy generation, demand, and grid operations. These systems facilitate efficient and coordinated integration of renewable energy sources in substations.

3. **Energy Storage Systems:** Energy storage systems, such as batteries and pumped hydro storage, can help mitigate the intermittency and variability of renewable energy sources. They store excess energy during periods of high generation and release it when demand exceeds generation.

4. **Flexible Generation and Demand Response:** Flexible generation resources, such as gas-fired power plants or hydroelectric facilities, can provide backup power and balance fluctuations in renewable energy generation. Demand response programs allow consumers to adjust their electricity usage in response to grid conditions, reducing the need for additional generation or grid reinforcement.

5. **Smart Grid Technologies:** Smart grid technologies, including advanced metering infrastructure, communication networks, and intelligent control systems, enable real-time monitoring, automation, and optimization of renewable energy integration. These technologies enhance grid flexibility, reliability, and efficiency.

6. **Microgrids and Islanded Systems:** In remote or off-grid areas, microgrids and islanded systems with renewable energy sources and energy storage can provide reliable and sustainable power supply.

Benefits of Renewable Energy Integration:

The integration of renewable energy sources in substations offers numerous benefits:

1. **Environmental Benefits:** Renewable energy sources help reduce greenhouse gas emissions and combat climate change, contributing to a cleaner and more sustainable environment.

2. **Energy Independence:** Diversifying the energy mix with renewable sources enhances energy independence by reducing dependence on imported fossil fuels.
3. **Cost Reduction:** Renewable energy costs have been declining, and integrating renewables can lead to cost savings over the long term. As technology advances and economies of scale are achieved, renewable energy generation becomes more competitive.
4. **Job Creation and Economic Growth:** The renewable energy sector creates jobs, stimulates local economic development, and attracts investment, fostering a green economy.
5. **Community Engagement:** The integration of renewable energy sources can engage local communities, allowing them to participate in renewable energy projects and benefit from the local generation of clean energy.

Conclusion:

The integration of renewable energy sources in substations is a pivotal step towards achieving a sustainable energy future. Despite challenges related to intermittency, variability, and grid stability, solutions such as grid codes, energy management systems, energy storage, and smart grid technologies enable the effective integration of renewable energy sources. The benefits of renewable energy integration, including environmental sustainability, energy independence, cost reduction, and economic growth, make it a compelling pathway for substations and the broader power system. By embracing renewable energy integration, substations contribute to a cleaner, greener, and more resilient energy landscape.

Smart Grid Technologies and Their Impact on Substations

Smart grid technologies play a transformative role in the modernization and optimization of substations. These advanced technologies enable intelligent monitoring, control, and communication within the power system, enhancing the efficiency, reliability, and sustainability of substations. This section provides an elaborate exploration of smart grid technologies and their impact on substations.

Advanced Metering Infrastructure (AMI):

Advanced Metering Infrastructure, also known as smart meters, enables real-time monitoring and two-way communication between utilities and consumers. Smart meters provide accurate and timely consumption data, enabling demand response programs, real-time pricing, and enhanced load management. In substations, smart meters facilitate precise measurement and monitoring of energy flow, voltage levels, and power quality, supporting efficient energy management and grid operation.

Distribution Automation:

Distribution Automation involves the integration of intelligent devices, sensors, and communication networks in the distribution grid. These technologies enable remote monitoring, control, and fault detection, enhancing the reliability and efficiency of distribution systems. In substations, distribution automation systems improve fault identification and isolation, reducing outage duration and improving system restoration. Automated switching devices and reclosers enable rapid fault recovery, minimizing disruption to customers.

Intelligent Electronic Devices (IEDs):

Intelligent Electronic Devices, such as protective relays and smart switches, are equipped with advanced computational capabilities and communication interfaces. IEDs enable real-time monitoring, control, and protection of substation equipment. They enhance system visibility, facilitate fault detection, and enable rapid fault response. IEDs communicate with substation control systems, exchange data, and provide valuable information for condition monitoring, predictive maintenance, and system optimization.

Communication Networks:

Communication networks form the backbone of smart grid technologies, enabling the exchange of data between substations, control centers, and other grid components. These networks utilize various communication technologies, such as fiber optics, wireless communication, and Internet Protocol (IP) networks. Communication networks enable real-time monitoring, control, and coordination of substations, supporting efficient energy management, fault detection, and rapid response to grid events.

Integrated Volt-VAR Control (IVVC):

Integrated Volt-VAR Control is a smart grid technology that optimizes voltage levels and reactive power flow in the distribution system. IVVC utilizes real-time data and advanced control algorithms to adjust voltage levels based on system conditions and load requirements. In substations, IVVC systems regulate voltage and reactive power at the distribution level, minimizing losses and enhancing system efficiency. IVVC also supports the integration of distributed energy resources (DERs) by managing voltage fluctuations caused by intermittent generation.

Demand Response (DR) Programs:

Demand Response programs engage consumers in managing their electricity usage in response to grid conditions. Through smart grid technologies, consumers receive real-time pricing signals, energy usage information, and control capabilities to adjust their electricity consumption. In substations, DR programs can help balance load, reduce peak demand, and enhance system reliability. By shifting or reducing energy consumption during periods of high demand, substations can alleviate strain on the grid and optimize generation and distribution resources.

Energy Management Systems (EMS):

Energy Management Systems provide centralized control, monitoring, and optimization of the entire power system. EMS integrates data from substations, generation units, and distribution networks to enable efficient grid operation. In substations, EMS facilitates real-time monitoring of equipment status, load profiles, and system performance. It supports decision-making processes for load dispatch, load forecasting, and optimal resource utilization. EMS also enables coordination with renewable energy sources, energy storage systems, and other distributed energy resources.

Grid Analytics and Big Data:

Grid Analytics and Big Data technologies leverage advanced data analytics algorithms to process large volumes of data collected from substations and the wider grid. These technologies provide valuable insights into grid performance, system vulnerabilities, and asset conditions. In substations, grid analytics enable predictive maintenance, asset health monitoring, and system optimization. By analyzing historical data, grid analytics can identify patterns, predict equipment failures, and optimize maintenance schedules, minimizing downtime and reducing operational costs.

Cybersecurity:

As substations become more interconnected and dependent on communication networks, cybersecurity becomes increasingly important. Smart grid technologies employ cybersecurity measures to protect substations from cyber threats, unauthorized access, and data breaches. These measures include encryption, firewalls, intrusion detection systems, and secure communication protocols. By safeguarding substations against cyber-attacks, the integrity, reliability, and safety of the power system are preserved.

Conclusion:

Smart grid technologies have a profound impact on substations, enhancing their efficiency, reliability, and sustainability. Advanced metering infrastructure, distribution automation, intelligent electronic devices, communication networks, IVVC, demand response programs, energy management systems, grid analytics, and cybersecurity all contribute to the transformation of substations into intelligent and interconnected entities. These technologies improve system visibility, enable real-time monitoring and control, support efficient energy management, optimize asset performance, and enhance grid resilience. By embracing smart grid technologies, substations can adapt to evolving energy landscapes, integrate renewable energy sources, and contribute to the development of a sustainable and resilient power system.

Advanced communication and control systems

Advanced communication and control systems play a crucial role in the modernization and optimization of substations. These systems enable efficient data exchange, real-time monitoring, control, and coordination of substation equipment, enhancing the reliability, safety, and performance of the power system. This section provides an elaborate exploration of advanced communication and control systems in substations.

Importance of Advanced Communication and Control Systems:

Advanced communication and control systems are essential for substations due to the following reasons:

1. **Enhanced Monitoring and Control:** These systems enable real-time monitoring of substation equipment, including transformers, circuit breakers, relays, and switches. They provide accurate and timely information on system conditions, enabling operators to make informed decisions and take appropriate actions.

2. **Improved Situational Awareness:** Advanced communication and control systems enhance situational awareness by collecting and integrating data from various sources within the substation. This includes data on equipment status, power quality, fault conditions, and grid conditions. Operators gain a comprehensive view of the substation, enabling them to identify potential issues and respond effectively.

3. **Remote Control and Automation:** These systems facilitate remote control and automation of substation equipment. Operators can remotely operate circuit breakers, switches, and other devices, reducing the need for manual intervention. Automation capabilities enable efficient and precise control of power flow, fault detection, and restoration processes.

4. **Faster Fault Detection and Response:** Advanced communication and control systems enable quick fault detection and response in substations. Real-time monitoring and data analysis enable the identification of abnormal conditions and prompt initiation of protection schemes. Operators can rapidly isolate faults, restore power, and minimize disruption to the power system.

5. **Coordination and Integration:** These systems enable coordination and integration of substation equipment with the wider power system. They facilitate communication between substations, control centers, and other grid

components, ensuring efficient operation, optimized power flow, and coordinated response to grid events.

Communication Technologies in Substations:

Advanced communication systems utilize various technologies for efficient data exchange and information sharing in substations. Some commonly used communication technologies include:

1. **Ethernet:** Ethernet-based communication networks provide high-speed, reliable, and secure data transmission within substations. Ethernet protocols, such as IEC 61850, enable seamless integration of substation devices, control systems, and data exchange.
2. **Fiber Optics:** Fiber optic cables offer high bandwidth, low latency, and immunity to electromagnetic interference. They provide fast and reliable communication links for transmitting large volumes of data between substation devices, control centers, and other grid components.
3. **Wireless Communication:** Wireless technologies, such as Wi-Fi, cellular networks, and radio frequency, enable flexible and cost-effective communication in substations. They provide connectivity for remote monitoring, control, and data exchange, eliminating the need for physical cabling.
4. **Power Line Communication (PLC):** PLC utilizes the existing power line infrastructure to transmit data signals. It enables communication between devices connected to the power grid, such as smart meters and substation equipment, using power cables as communication channels.
5. **Satellite Communication:** Satellite communication provides connectivity in remote areas or areas with limited terrestrial network coverage. It ensures reliable and continuous data exchange between substations and control centers, enabling efficient monitoring and control.

Control Systems and SCADA:

Supervisory Control and Data Acquisition (SCADA) systems form the backbone of substation control and monitoring. SCADA systems collect data from substation devices, perform real-time monitoring, control equipment operation, and provide operators with a graphical interface for system visualization and control. Advanced SCADA systems offer the following features:

1. **Real-time Monitoring:** SCADA systems continuously monitor substation equipment, power parameters, and grid conditions. Operators can view real-time data on voltage levels, current flows, power quality, and alarm conditions, enabling quick response to abnormal events.

2. **Remote Control:** SCADA systems enable remote control of substation equipment, such as circuit breakers, switches, and tap changers. Operators can initiate control actions, switch between operating modes, and remotely operate devices for maintenance purposes.

3. **Data Logging and Analysis:** SCADA systems collect and store historical data, allowing operators to analyze trends, identify patterns, and perform diagnostics. Data analysis supports condition monitoring, predictive maintenance, and optimization of substation performance.

4. **Alarm and Event Management:** SCADA systems provide alarms and event notifications based on predefined thresholds or abnormal conditions. Operators receive timely alerts, enabling them to take immediate action to prevent equipment failures or mitigate potential grid disturbances.

5. **Integration with Protection Systems:** SCADA systems integrate with substation protection systems, enabling coordinated protection schemes. They facilitate

communication between protection relays, tripping devices, and control systems, ensuring effective fault detection, isolation, and restoration.

Intelligent Electronic Devices (IEDs):

Intelligent Electronic Devices (IEDs) are key components of advanced control systems in substations. These devices, such as protective relays, smart meters, and intelligent switches, offer advanced functionality, communication capabilities, and computational power. IEDs perform functions such as fault detection, metering, data acquisition, control, and communication. They exchange data with SCADA systems, protection relays, and other devices, enabling efficient and coordinated substation operation.

Integration with Energy Management Systems (EMS) and Distributed Energy Resources (DERs):

Advanced communication and control systems integrate with Energy Management Systems (EMS) and Distributed Energy Resources (DERs). EMS provides centralized control, optimization, and coordination of generation, transmission, and distribution assets. Integration with EMS enables efficient load dispatch, power system stability, and integration of renewable energy sources. Communication and control systems also facilitate the monitoring, control, and optimization of DERs, such as solar PV systems, wind turbines, and energy storage devices, within substations.

Cybersecurity Considerations:

As communication and control systems become more interconnected, ensuring cybersecurity becomes vital. Substations are vulnerable to cyber threats, including unauthorized access, data breaches, and malicious attacks. Robust cybersecurity measures, such as encryption, firewalls, intrusion detection systems, and access control mechanisms, must be implemented to safeguard substations from cyber threats and maintain the integrity and reliability of the power system.

Conclusion:

Advanced communication and control systems are integral to the modernization and optimization of substations. These systems enable real-time monitoring, control, and coordination of substation equipment, enhancing system reliability, efficiency, and safety. Through technologies such as SCADA, advanced communication networks, IEDs, and integration with EMS and DERs, substations become intelligent, interconnected entities within the power grid. By embracing advanced communication and control systems, substations contribute to the development of a smart grid that supports efficient energy management, optimal power flow, integration of renewable energy sources, and improved grid resilience.

Cybersecurity Considerations in Substations

Cybersecurity is of utmost importance in substations as they become increasingly connected and reliant on digital technologies. Substations are vulnerable to cyber threats that can compromise the integrity, availability, and confidentiality of critical operational systems. This section provides an elaborate exploration of cybersecurity considerations in substations, including the challenges, best practices, and technologies employed to ensure a secure substation environment.

Importance of Substation Cybersecurity:

Substation cybersecurity is crucial due to the following reasons:

1. **Protecting Critical Infrastructure:** Substations are critical components of the power grid, and any compromise to

their operation can have severe consequences on the reliability and safety of the entire electricity supply chain.

2. **Safeguarding Operational Integrity:** Cybersecurity measures in substations ensure the integrity of operational systems, including protection relays, control systems, SCADA, and communication networks. Protection relays, for example, play a vital role in fault detection, isolation, and system restoration. Compromising these systems can lead to widespread disruptions or even physical damage.

3. **Maintaining Grid Reliability:** Substation cybersecurity is essential to maintain grid reliability by preventing unauthorized access, data manipulation, or disruption of critical operational functions. Unintended outages, misoperation, or malicious attacks can have cascading effects on the stability of the power system.

4. **Ensuring Data Confidentiality:** Substations generate and exchange sensitive data related to grid operation, customer information, and system vulnerabilities. Protecting this data from unauthorized access or disclosure is essential to maintain customer privacy, comply with regulations, and safeguard critical information.

Cybersecurity Challenges in Substations:

Substations face unique cybersecurity challenges that require specific attention:

1. **Legacy Systems:** Many substations have legacy control systems and devices with limited built-in security features. These systems may lack regular software updates or patches, making them vulnerable to known security vulnerabilities.

2. **Interconnected Environment:** Substations are increasingly interconnected with control centers, remote access

systems, and other grid components. The interconnected nature of substations increases the attack surface and exposes them to potential cyber threats.

3. **Third-Party Risks:** Substations often rely on third-party vendors for equipment, software, and services. These dependencies introduce additional cybersecurity risks, such as supply chain attacks or vulnerabilities in vendor-provided systems.
4. **Human Factors:** Human error, including unintentional actions or lack of cybersecurity awareness, can pose significant risks. Phishing attacks, social engineering, or improper handling of security protocols by employees can compromise the security of substations.
5. **Evolving Threat Landscape:** Cyber threats continuously evolve, and attackers are becoming more sophisticated in their methods. Substations must adapt and stay ahead of emerging threats to protect against cyber attacks.

Best Practices for Substation Cybersecurity:

To address these challenges, several best practices should be followed to ensure robust cybersecurity in substations:

1. **Risk Assessment:** Conducting a comprehensive risk assessment to identify potential vulnerabilities, threats, and impacts is crucial. This assessment should include an evaluation of physical security, network architecture, access controls, and potential attack vectors.
2. **Security Policies and Procedures:** Developing and implementing cybersecurity policies and procedures is essential to establish a strong security posture. These policies should address aspects such as user access management, password policies, incident response

protocols, and employee training on cybersecurity awareness.

3. **Defense-in-Depth Strategy:** Implementing a defense-in-depth strategy involves employing multiple layers of security controls to protect substations. This includes measures such as firewalls, intrusion detection systems, access controls, encryption, and regular patching of software and firmware.

4. **Network Segmentation:** Segmenting substation networks into isolated zones with controlled access helps minimize the impact of a security breach. Critical devices and systems should be isolated from less critical or external networks to contain potential attacks and limit unauthorized access.

5. **Secure Remote Access:** Implementing secure remote access mechanisms for authorized personnel is essential for maintenance and monitoring. This involves strong authentication, encrypted communication channels, and access control mechanisms to prevent unauthorized access.

6. **Employee Awareness and Training:** Conducting regular training programs to educate employees about cybersecurity risks, best practices, and incident response protocols is crucial. Employees should be aware of phishing attacks, social engineering techniques, and the importance of adhering to security policies.

7. **Incident Response and Recovery:** Establishing an effective incident response plan helps to quickly identify, contain, and mitigate potential cyber incidents. This plan should include procedures for reporting incidents, isolating affected systems, restoring operations, and conducting post-incident analysis.

Cybersecurity Technologies and Solutions:

Various technologies and solutions can enhance cybersecurity in substations:

1. **Intrusion Detection Systems (IDS) and Intrusion Prevention Systems (IPS):** IDS and IPS monitor network traffic for suspicious activities and can automatically block or alert administrators about potential cyber threats.
2. **Secure Communication Protocols:** Implementing secure communication protocols, such as secure VPNs (Virtual Private Networks) or encrypted communication channels, ensures the confidentiality and integrity of data transmitted within and between substations.
3. **Security Information and Event Management (SIEM) Systems:** SIEM systems aggregate, correlate, and analyze security event logs from multiple devices and systems. They help detect anomalies, identify potential attacks, and facilitate timely incident response.
4. **Security Patch Management:** Regularly applying security patches and updates to substation software, firmware, and devices is critical to address known vulnerabilities and protect against exploits.
5. **Security Monitoring and Auditing:** Continuous security monitoring and auditing of substation systems and networks can help detect potential breaches, unusual activities, or unauthorized access attempts.
6. **Encryption and Data Protection:** Encrypting sensitive data at rest and in transit provides an additional layer of protection against unauthorized access and data manipulation.
7. **Physical Security Measures:** Implementing physical security measures, such as access controls, video surveillance, and intrusion detection systems, helps prevent unauthorized physical access to substations and critical equipment.

Compliance and Regulations:

Substations must comply with relevant cybersecurity regulations and standards. These may include industry-specific standards like IEC 62443 or regional regulations such as NERC CIP (North American Electric Reliability Corporation Critical Infrastructure Protection). Compliance with these standards helps ensure a minimum level of cybersecurity and provides a framework for risk assessment, mitigation, and incident response.

Conclusion:

Cybersecurity considerations are paramount in substations to protect critical infrastructure, maintain operational integrity, and ensure the reliability of the power grid. By implementing robust cybersecurity measures, following best practices, and leveraging advanced technologies, substations can safeguard against cyber threats and mitigate potential risks. Continuous monitoring, regular risk assessments, employee training, and adherence to cybersecurity policies contribute to creating a secure substation environment in the face of evolving cyber threats.

CHAPTER 8: CASE STUDIES

Chapter 8 presents a collection of case studies that showcase real-world examples of electrical substations. These case studies highlight various aspects of substation design, operation, and challenges faced in different scenarios. By exploring these cases, readers can gain practical insights into the implementation of substations and the solutions employed to address specific requirements or issues. This chapter aims to provide valuable lessons and inspiration for substation professionals and those interested in the field.

Chapter 8 provides a diverse collection of case studies that showcase real-world examples of electrical substations. These case studies shed light on different aspects of substation design, operation, and challenges, offering valuable insights and lessons for professionals in the field. By examining these real-world scenarios, readers can gain practical knowledge and inspiration for addressing specific requirements, implementing innovative solutions, and ensuring the reliable and efficient operation of electrical substations.

Real-world examples of substation design, implementation, and challenges

Real-world examples of substation design, implementation, and challenges provide valuable insights into the complexities and considerations involved in building and operating electrical substations. This section explores a range of real-world examples that highlight various aspects of substation projects, the challenges faced, and the solutions employed.

Example 1: Large-Scale Transmission Substation:

A large-scale transmission substation serves as a critical hub for transmitting high-voltage power across long distances. These substations often connect multiple transmission lines and interconnect different parts of the power grid. Challenges in designing and implementing such substations include the need for high-capacity transformers, advanced protection systems, and robust communication networks. The design must account for factors such as fault tolerance, redundancy, and grid stability. Solutions typically involve the integration of advanced monitoring and control systems, flexible power flow management, and coordination with regional transmission operators.

Example 2: Urban Distribution Substation:

Distribution substations serve the purpose of stepping down voltage levels for local distribution to end consumers. In urban areas, these substations face challenges related to limited space availability, environmental considerations, and aesthetic integration with the surroundings. Designing compact and aesthetically pleasing substations while meeting electrical and safety requirements is crucial. Solutions may involve the use of gas-insulated switchgear (GIS) to reduce the substation's footprint, underground cabling, and architectural designs that blend with the urban landscape. Noise reduction measures and community engagement are also important considerations.

Example 3: Renewable Energy Integration Substation:

Substations dedicated to integrating renewable energy sources, such as solar or wind farms, present unique challenges. These substations must handle intermittent and variable power generation, grid stability concerns, and bidirectional power flow. Design considerations include the capacity to manage fluctuations in renewable energy output, accommodate multiple energy sources, and enable smooth integration with the existing power grid. Solutions involve advanced control systems, energy storage integration, and dynamic grid management techniques to optimize power flow and ensure grid stability.

Example 4: Remote Off-Grid Substation:

Remote off-grid substations serve areas without access to the main power grid. These substations often rely on localized power generation, such as diesel generators or renewable energy sources, combined with energy storage systems. Challenges include ensuring reliable power supply, managing energy storage, and optimizing power generation based on demand. Solutions may involve advanced microgrid control systems, predictive load forecasting, and hybrid power generation technologies. Energy efficiency and sustainability are key considerations in these off-grid substations.

Example 5: Substation Retrofit and Modernization:

Existing substations often require retrofitting and modernization to accommodate growing demand, technology upgrades, or compliance with new regulations. Challenges include maintaining operational continuity during the retrofit process, integrating new technologies with legacy equipment, and minimizing downtime. Solutions may involve phased upgrades, temporary parallel systems, and careful planning to ensure a smooth transition. Advanced monitoring and control systems, including digital relays, intelligent switches, and SCADA systems, are commonly integrated during the modernization process.

Example 6: Substation Expansion in Brownfield Sites:

Expanding substations in existing brownfield sites presents challenges due to limited space availability and potential disruption to ongoing operations. The design must optimize space utilization, ensure safety during construction, and minimize impacts on the surrounding environment. Solutions may involve compact equipment configurations, innovative cable routing, and temporary power supply arrangements. Efficient project management and coordination are essential to complete the expansion while maintaining grid reliability.

Example 7: Substation Rehabilitation and Refurbishment:

Aged substations often require rehabilitation and refurbishment to ensure reliability, safety, and compliance with modern standards. Challenges include identifying equipment requiring replacement or refurbishment, managing the logistical aspects of the project, and addressing potential risks associated with outdated systems. Solutions may involve condition assessment, selective replacement of components, equipment refurbishment, and adopting new technologies for improved performance. Retrofitting with advanced protection systems, control systems, and monitoring equipment may be necessary.

Conclusion:

Real-world examples of substation design, implementation, and challenges provide practical insights into the complexities of building and operating electrical substations. These examples demonstrate the diverse range of projects and the considerations required to meet specific requirements and overcome various challenges. By studying these real-world cases, professionals in the field can gain valuable knowledge and inspiration to address similar challenges in their own substation projects, ensuring the reliable, efficient, and sustainable operation of electrical substations.

Case Studies of Substation Upgrades, Retrofits, and Expansions

Substation upgrades, retrofits, and expansions are essential to meet growing energy demands, integrate new technologies, enhance reliability, and comply with evolving industry standards. This section presents a collection of case studies that highlight real-world examples of substation projects involving upgrades, retrofits, and expansions. These case studies provide insights into the challenges faced, the solutions implemented, and the benefits achieved through these projects.

Case Study 1: Substation Upgrade for Increased Capacity:

In this case study, a substation required an upgrade to increase its capacity and accommodate the growing electricity demand in the area. The project involved replacing aging equipment, such as transformers and circuit breakers, with higher-capacity units. The challenges included coordinating the replacement activities without interrupting the power supply and ensuring compatibility between the new and existing equipment. The solution involved careful planning, staged equipment replacement, and temporary parallel operations to maintain power availability during the upgrade. The result was an expanded substation capable of meeting the increased load demand while improving system reliability.

Case Study 2: Retrofitting Protection and Control Systems:

This case study focuses on retrofitting the protection and control systems of an existing substation with advanced digital relays, automation devices, and communication systems. The goal was to enhance the substation's operational efficiency, system monitoring capabilities, and fault detection. The challenges involved integrating new technologies with the legacy infrastructure and ensuring seamless interoperability between the upgraded and existing systems. The solution included a phased approach, thorough testing, and coordination with the existing equipment. The retrofit resulted in improved system reliability, faster fault detection, and enhanced situational awareness for operators.

Case Study 3: Expansion of Distribution Substation:

This case study explores the expansion of a distribution substation to meet the increasing electricity demand in a rapidly growing urban area. The project involved adding new transformers, switchgear, and distribution feeders to expand the substation's capacity. Challenges included limited available space and the need to minimize disruptions to the existing distribution network. The solution involved careful site planning, compact equipment designs, and temporary power supply arrangements during construction. The expanded substation improved power reliability, reduced distribution losses, and ensured adequate supply to the growing consumer base.

Case Study 4: Substation Automation Retrofit:

This case study focuses on the retrofit of an existing substation with advanced automation and control systems. The objective was to improve the substation's remote monitoring and control capabilities, enhance operational efficiency, and enable integration with the utility's SCADA system. The challenges included integrating the new automation systems with the existing substation infrastructure and ensuring reliable communication between devices. The solution involved the installation of intelligent electronic devices (IEDs), communication networks, and SCADA integration. The retrofit provided real-time data access, enhanced fault detection, and improved response times, resulting in more efficient substation operation and maintenance.

Case Study 5: Substation Refurbishment for Environmental Compliance:

In this case study, an aging substation required refurbishment to meet new environmental regulations and reduce its environmental impact. The project involved the replacement of oil-filled equipment with modern, environmentally friendly alternatives, such as gas-insulated switchgear (GIS) and dry-type transformers. The challenges included minimizing downtime during the retrofit and ensuring compatibility between the new and existing equipment. The solution included meticulous planning, temporary power supply arrangements, and proper disposal of the old equipment. The refurbished substation achieved compliance with environmental regulations, reduced environmental risks, and improved overall substation performance.

Case Study 6: Grid Modernization and Renewable Integration:

This case study focuses on a substation upgrade and retrofit to facilitate the integration of renewable energy sources into the grid. The project involved the installation of advanced control systems, grid management technologies, and energy storage systems. Challenges included managing the intermittency of renewable energy generation, optimizing power flow, and ensuring grid stability. The solution included the implementation of intelligent grid management algorithms, enhanced protection schemes, and coordination with renewable energy installations. The upgraded substation enabled the efficient integration of renewable energy, improved grid stability, and increased renewable energy penetration levels.

Case Study 7: Substation Expansion in a Brownfield Site:

In this case study, an existing substation situated in a constrained brownfield site required expansion to meet increased demand and accommodate new equipment. The project faced challenges such as limited available space, complex interconnections, and the need to maintain ongoing substation operations. The solution involved careful site planning, compact equipment designs, and staged construction to minimize disruptions. Advanced cable routing techniques and temporary power supply arrangements were implemented during the expansion. The expanded substation provided additional capacity, improved system reliability, and facilitated future equipment upgrades.

Conclusion:

Case studies of substation upgrades, retrofits, and expansions provide practical insights into the challenges and solutions encountered in real-world projects. These examples demonstrate the importance of careful planning, phased implementation, compatibility assessments, and coordination to ensure successful substation modifications. By studying these cases, professionals in the field can gain valuable knowledge and inspiration for similar projects, enabling them to optimize substation performance, improve reliability, and meet the evolving needs of the electrical power system.

Lessons Learned and Best Practices from Successful Substation Projects

Successful substation projects provide valuable lessons and best practices that can be applied to future endeavors. This section explores some key lessons learned and best practices derived from successful substation projects, highlighting the factors that contribute to project success, efficiency, and reliability.

Thorough Planning and Project Management:

Lesson: Thorough planning is essential for project success. A well-defined scope, clear objectives, and a comprehensive project plan are crucial.

Best Practice: Conduct a detailed feasibility study, risk assessment, and stakeholder analysis to identify project requirements, potential challenges, and key stakeholders. Develop a project management plan with defined roles and responsibilities, timelines, and resource allocation. Regularly monitor and track progress to ensure project milestones are met.

Collaboration and Communication:

Lesson: Effective collaboration and communication among all project stakeholders are vital for project success.

Best Practice: Foster open communication channels between the project team, contractors, vendors, and other relevant parties. Hold regular meetings, encourage feedback, and address issues promptly. Implement project management tools and systems that facilitate efficient information sharing and collaboration.

Comprehensive Design and Engineering:

Lesson: A comprehensive and detailed design is critical to ensuring efficient and reliable substation operation.

Best Practice: Engage experienced design and engineering professionals who have a thorough understanding of substation requirements and applicable standards. Conduct detailed site surveys, analyze load requirements, and account for future growth and technology advancements. Utilize advanced design tools and simulation software to optimize the substation layout, equipment placement, and electrical system design.

Robust Equipment Selection:

Lesson: Selecting high-quality and reliable equipment is crucial for long-term substation performance and reliability.

Best Practice: Conduct a thorough evaluation of equipment manufacturers, considering factors such as track record, reputation, quality certifications, and compliance with industry standards. Assess the suitability of equipment for the specific project requirements, considering factors like capacity, reliability, compatibility, and maintenance requirements. Consider the lifecycle cost, including maintenance, operational, and replacement costs, when making equipment selection decisions.

Safety Considerations:

Lesson: Safety should be a top priority throughout the project lifecycle, from design to construction and operation.

Best Practice: Incorporate safety considerations into every aspect of the project, including design, equipment selection, construction, and maintenance. Comply with relevant safety regulations and standards. Conduct regular safety training for project personnel and ensure adherence to safety protocols and procedures. Implement appropriate safety measures such as signage, personal protective equipment, and safety barriers during construction and operation.

Quality Control and Testing:

Lesson: Stringent quality control measures and comprehensive testing are essential for ensuring the reliability and performance of substation equipment.

Best Practice: Implement a quality control plan that includes regular inspections, factory acceptance tests, and on-site testing. Engage qualified testing personnel and equipment to conduct thorough checks and verifications. Test equipment and systems under various operating conditions to identify any potential issues or performance deviations. Document and review test results to ensure compliance with design specifications and standards.

Robust Commissioning and Handover Process:

Lesson: A well-planned commissioning and handover process is crucial for a smooth transition from construction to operational phase.

Best Practice: Develop a comprehensive commissioning plan that includes pre-commissioning checks, equipment energization, functional testing, and performance verification. Conduct thorough inspections, review documentation, and ensure that all systems and equipment are operating as intended. Train operational staff on the operation, maintenance, and safety procedures of the substation. Ensure proper documentation of as-built drawings, equipment manuals, and maintenance schedules for future reference.

Continuous Monitoring and Maintenance:

Lesson: Continuous monitoring and maintenance are vital for the long-term reliability and performance of the substation.

Best Practice: Implement a proactive maintenance program that includes regular inspections, equipment testing, and preventive maintenance activities. Utilize advanced monitoring systems and diagnostic tools to identify potential issues and take preventive measures. Regularly review and update maintenance procedures and schedules based on equipment performance and operational experience. Promote a culture of continuous improvement and knowledge sharing among operational staff.

Documentation and Knowledge Management:

Lesson: Proper documentation and knowledge management facilitate efficient operation, maintenance, and future upgrades of the substation.

Best Practice: Maintain accurate and up-to-date documentation, including design drawings, equipment manuals, test reports, and maintenance records. Establish a centralized information repository for easy access to relevant project documents and operational data. Promote knowledge sharing among project team members and operational staff through training programs, workshops, and knowledge transfer activities.

Conclusion:

Lessons learned and best practices from successful substation projects contribute to the ongoing improvement and optimization of future endeavors. Thorough planning, effective collaboration, comprehensive design, robust equipment selection, safety considerations, quality control, and continuous monitoring are critical factors for successful substation projects. By applying these lessons and best practices, stakeholders can enhance project efficiency, reliability, and the long-term performance of electrical substations.

CHAPTER 9: SAFETY, REGULATIONS, AND ENVIRONMENTAL IMPACT

Chapter 9 focuses on the critical aspects of safety, regulations, and environmental impact associated with electrical substations. Safety measures, adherence to regulations, and consideration of environmental impact are crucial for the reliable and sustainable operation of substations. This chapter explores the key elements of safety practices, regulatory compliance, and environmental considerations in the design, construction, and operation of substations.

Chapter 9 emphasizes the importance of safety, regulatory compliance, and environmental impact considerations in the design, construction, and operation of electrical substations. Safety measures and practices ensure the well-being of personnel and mitigate risks associated with high-voltage equipment. Regulatory compliance ensures adherence to standards, codes, and regulations governing substations, enabling seamless integration with the power grid. Environmental considerations help minimize the impact of substations on the surrounding ecosystem and contribute to sustainable energy infrastructure. By prioritizing safety, adhering to regulations, and considering environmental impacts, substations can operate reliably, safely, and sustainably in harmony with their surroundings.

Safety Protocols and Guidelines for Substation Personnel

Safety protocols and guidelines are crucial for ensuring the well-being of substation personnel and mitigating potential hazards associated with working in high-voltage environments. This section elaborates on key safety protocols and guidelines that should be followed by substation personnel to maintain a safe working environment.

Personal Protective Equipment (PPE):

- Substation personnel should wear appropriate PPE to protect themselves from electrical, thermal, and physical hazards.
- PPE may include safety helmets, safety glasses, ear protection, flame-resistant clothing, gloves, and safety footwear.

- PPE should be well-maintained, properly fitted, and in compliance with relevant safety standards.

Lockout/Tagout (LOTO) Procedures:

- LOTO procedures are essential for ensuring the safety of personnel during equipment maintenance or repair.
- Personnel must follow established LOTO procedures to de-energize equipment, isolate energy sources, and apply lockout/tagout devices to prevent accidental energization.
- LOTO procedures should be clearly documented, regularly reviewed, and strictly followed to prevent unauthorized access and accidental re-energization.

Electrical Hazard Awareness and Mitigation:

- Substation personnel must have a thorough understanding of electrical hazards and the safe work practices to mitigate these risks.
- Proper training should be provided on identifying electrical hazards, working on energized equipment (when necessary), and understanding the use of safety equipment and tools.
- Precautions must be taken to avoid electrical shocks, arc flashes, and contact with live parts, including maintaining safe working distances, using insulated tools, and employing proper grounding techniques.

Fire Prevention and Firefighting Systems:

- Substations should have adequate fire prevention measures and firefighting systems in place.

- Personnel should be trained in fire prevention techniques, including proper storage and handling of flammable materials and regular inspection of electrical connections.
- Firefighting systems, such as fire extinguishers, fire suppression systems, and emergency response procedures, should be readily available and well-maintained.

Emergency Response Planning and Drills:

- Substation personnel should be familiar with emergency response plans and procedures.
- Regular emergency drills and training exercises should be conducted to ensure personnel are prepared to respond quickly and effectively in case of emergencies, such as equipment failures, fires, or electrical faults.
- Emergency exit routes, assembly points, and communication protocols should be clearly marked and regularly reviewed.

Training and Education:

- Comprehensive training programs should be provided to substation personnel on safety protocols, hazard identification, and emergency response procedures.
- Training should cover topics such as electrical safety, first aid, CPR (cardiopulmonary resuscitation), and the use of fire extinguishers.
- Ongoing safety education programs and refresher courses should be conducted to keep personnel updated on the latest safety practices and technologies.

Reporting and Incident Investigation:

- Substation personnel should be encouraged to report any safety concerns, near misses, or accidents promptly.
- Incidents and near misses should be thoroughly investigated to identify root causes and implement corrective measures.
- Lessons learned from incidents should be shared among personnel to enhance safety awareness and prevent future occurrences.

Safety Culture and Continuous Improvement:

- A strong safety culture should be fostered within the substation environment.
- Management should lead by example and prioritize safety, promoting a proactive and collaborative approach to safety.
- Regular safety audits, inspections, and reviews should be conducted to identify potential safety hazards and areas for improvement.
- Feedback from substation personnel should be encouraged to continuously enhance safety protocols and guidelines.

Conclusion:

Safety protocols and guidelines are vital for substation personnel to ensure their well-being and mitigate hazards associated with working in high-voltage environments. By following these protocols, wearing appropriate PPE, adhering to LOTO procedures, being aware of electrical hazards, and actively participating in safety training and drills, substation personnel can create a safe working environment. Continuous improvement, a strong safety culture, and regular safety audits further contribute to maintaining a safe substation environment.

Regulatory Compliance and Industry Standards

Regulatory compliance and adherence to industry standards are crucial for the safe, efficient, and reliable operation of electrical substations. This section provides an elaboration on the importance of regulatory compliance and the key industry standards that govern the design, construction, operation, and maintenance of substations.

Importance of Regulatory Compliance:

- Regulatory compliance ensures that substations meet legal requirements and operate in accordance with established guidelines.
- Compliance helps safeguard the health and safety of personnel, protect the environment, and maintain grid stability.
- Non-compliance can result in penalties, legal liabilities, and reputational damage.

Regulatory Bodies:

Regulatory bodies at various levels, such as national, regional, and local authorities, oversee the electrical power industry and establish regulations. These bodies develop and enforce regulations to ensure the reliability, safety, and interoperability of electrical substations.

Regulatory Areas:

Regulatory requirements cover a wide range of aspects related to substations, including design, construction, equipment standards, operation, and maintenance. Common regulatory areas include safety regulations, grid codes, environmental regulations, and equipment certification.

Safety Regulations:

Safety regulations focus on protecting personnel, property, and the public from electrical hazards. They specify safety practices, personnel training requirements, safety equipment standards, and emergency response procedures.

Safety regulations ensure compliance with international standards, such as the International Electrotechnical Commission (IEC) 61850, Occupational Safety and Health Administration (OSHA) standards, and local safety regulations.

Grid Codes and Interconnection Standards:

Grid codes define the technical requirements for connecting substations to the power grid. They specify voltage levels, frequency limits, power quality standards, and operational procedures to ensure seamless integration with the grid. Compliance with grid codes is essential to maintain grid stability, support power system operation, and facilitate reliable power transmission and distribution.

Environmental Regulations:

Environmental regulations govern the impact of substations on the environment, including air quality, water pollution, noise levels, and habitat preservation. They require substations to comply with emission standards, waste management practices, and environmental impact assessments. Compliance with environmental regulations ensures sustainable operation, minimizes ecological impact, and supports responsible energy infrastructure development.

Equipment Standards and Certification:

Equipment standards ensure that substation components meet specific performance, safety, and interoperability requirements. Equipment certification, such as the IEC standards, demonstrates that the equipment has undergone rigorous testing and complies with international standards. Compliance with equipment standards and certification helps ensure the reliability and compatibility of substation equipment.

Industry Standards Organizations:

Various organizations develop industry standards for substations, including the IEC, the National Electrical Manufacturers Association (NEMA), and the Institute of Electrical and Electronics Engineers (IEEE). These organizations publish standards related to equipment design, communication protocols, interoperability, and safety practices.

Compliance with industry standards helps maintain consistency, promotes interoperability, and ensures best practices in substation design and operation.

Compliance Monitoring and Enforcement:

Regulatory bodies monitor compliance through inspections, audits, and periodic assessments. Non-compliance may result in penalties, fines, license revocation, or project delays. Effective compliance monitoring and enforcement mechanisms ensure that substations adhere to regulations and industry standards.

Conclusion:

Regulatory compliance and adherence to industry standards are essential for the safe and reliable operation of electrical substations. Compliance with safety regulations, grid codes, environmental regulations, and equipment standards ensures the protection of personnel, maintains grid stability, minimizes environmental impact, and supports the interoperability of substations. Substation professionals should stay updated on relevant regulations and standards, engage in continuous professional development, and collaborate with regulatory bodies to ensure compliance and promote a culture of safety and quality within the industry.

Environmental Impact Assessment and Mitigation Measures

Environmental impact assessment (EIA) is a critical process that evaluates the potential environmental effects of a proposed substation project. This section elaborates on the importance of EIA and explores key mitigation measures to minimize the environmental impact of substations.

Importance of Environmental Impact Assessment:

Environmental Impact Assessment is a systematic process that identifies, predicts, and evaluates potential environmental impacts associated with a substation project. EIA ensures that the project is designed, constructed, and operated in an environmentally responsible manner. It considers the ecological, social, and economic aspects to achieve sustainable development.

EIA Process:

The EIA process typically involves the following steps:

- Scoping: Identifying the scope of the assessment, stakeholders, and potential environmental impacts.
- Baseline Study: Conducting an assessment of the existing environmental conditions in the project area.
- Impact Assessment: Identifying and evaluating potential impacts on various environmental components, such as air, water, soil, biodiversity, and socio-economic factors.

- Mitigation Measures: Developing strategies and measures to minimize, avoid, or mitigate identified impacts.
- Environmental Management Plan: Outlining measures for monitoring, mitigation, and management throughout the project lifecycle.
- Public Consultation: Engaging with stakeholders and seeking their input to ensure transparency and accountability.

Key Environmental Impact Areas:

1. **Air Quality:** Substations may generate air emissions during operations. The EIA assesses potential impacts on air quality, including emissions of pollutants such as sulfur dioxide, nitrogen oxides, and particulate matter.
2. **Water Resources:** EIA evaluates potential impacts on surface water and groundwater quality, as well as the potential for water pollution from construction activities and operational discharges.
3. **Soil and Land:** Impacts on soil erosion, land degradation, and soil contamination are assessed, considering potential disturbances during construction and the potential release of contaminants.
4. **Biodiversity and Ecological Systems:** The EIA evaluates potential impacts on habitats, vegetation, wildlife, and protected species in the project area. It assesses the potential for habitat fragmentation and loss of biodiversity.
5. **Noise and Vibration:** EIA assesses the potential impacts of noise and vibration from substation operations on nearby communities and sensitive receptors.
6. **Visual Impact:** The visual impact assessment considers the aesthetics of the substation and its integration into the surrounding landscape.

Mitigation Measures:

Environmental mitigation measures aim to minimize or eliminate potential adverse impacts identified during the EIA process. These measures may include:

- Implementing erosion and sediment control measures during construction to prevent soil erosion and water pollution.
- Installing noise barriers and utilizing low-noise equipment to mitigate noise impacts on surrounding communities.
- Implementing best management practices for waste management, including proper disposal and recycling of construction and operational waste.
- Incorporating landscaping and vegetation buffers to minimize visual impacts and enhance the integration of the substation into the surroundings.
- Implementing stormwater management strategies to control runoff and prevent water pollution.
- Adopting energy-efficient design and operational practices to minimize energy consumption and greenhouse gas emissions.
- Implementing measures to protect and enhance biodiversity, such as habitat restoration or creation programs.

Environmental Management Plan:

An Environmental Management Plan (EMP) outlines the measures to be implemented during the construction, operation, and decommissioning phases of the substation. The EMP includes monitoring programs, reporting mechanisms, and contingency plans to address any unforeseen environmental issues. The EMP ensures ongoing compliance with environmental regulations and commitment to sustainable practices throughout the project lifecycle.

Compliance and Monitoring:

Compliance with environmental regulations and commitments outlined in the EIA and EMP is essential. Regular monitoring and reporting are conducted to assess the effectiveness of mitigation measures and ensure ongoing compliance. Environmental audits and inspections may be performed to verify compliance and identify opportunities for improvement.

Conclusion:

Environmental Impact Assessment and the implementation of appropriate mitigation measures are essential for minimizing the environmental impact of substations. By conducting a comprehensive EIA, identifying potential impacts, and implementing effective mitigation measures, substations can be designed, constructed, and operated in an environmentally responsible manner. Compliance with environmental regulations and ongoing monitoring ensure that substations contribute to sustainable development and minimize their ecological footprint.

CHAPTER 10: FUTURE OUTLOOK AND CONCLUSION

Chapter 10 provides a glimpse into the future outlook of electrical substations, highlighting emerging trends, advancements in technology, and the challenges and opportunities that lie ahead. It concludes the book by summarizing the key points discussed throughout the chapters and emphasizing the importance of electrical substations in the evolving power sector.

Anticipated Developments in Substation Technology and Infrastructure

The field of substation technology and infrastructure is undergoing significant advancements driven by the need for more reliable, efficient, and sustainable power systems. This section elaborates on the anticipated developments in substation technology and infrastructure that will shape the future of electrical substations.

Digitalization and Intelligent Substations:

The adoption of digital technologies and intelligent systems is transforming substations into smart and interconnected entities. Advanced sensors, communication networks, and data analytics enable real-time monitoring, diagnostics, and predictive maintenance, improving substation performance and reducing downtime. Intelligent substations facilitate remote control, automation, and optimization of power flow, enabling faster fault detection, response, and restoration.

Flexible AC Transmission Systems (FACTS):

FACTS devices, such as Static Var Compensators (SVC) and Flexible AC Transmission System (FACTS) controllers, enhance power system stability, voltage control, and reactive power compensation. FACTS devices installed within substations improve power quality, mitigate transmission line congestion, and enable efficient power transfer, resulting in a more resilient and flexible power grid.

HVDC and Power Electronics:

High Voltage Direct Current (HVDC) technology is increasingly being used for long-distance power transmission and interconnecting different regional grids.

HVDC converter stations within substations enable efficient power transmission, integration of renewable energy sources, and interconnection of asynchronous AC grids.

Power electronic devices, such as Voltage Source Converters (VSC), provide enhanced control and stability, supporting power flow control, voltage regulation, and grid synchronization.

Energy Storage Systems (ESS):

Energy storage systems, such as battery energy storage and flywheel systems, are gaining prominence in substations. ESS installed within substations provide peak load shaving, grid stabilization, and backup power capabilities, enhancing the reliability and resilience of the power system.

Energy storage also enables the integration of intermittent renewable energy sources by storing excess energy during low demand periods and releasing it during high demand or when renewable generation is low.

Grid-Forming Inverters and Microgrids:

Grid-forming inverters enable the operation of microgrids, which are small-scale, localized power systems. Substations can incorporate grid-forming inverters and microgrid control systems to facilitate the integration of distributed energy resources (DERs), such as solar panels and wind turbines, enabling islanding capability and supporting local power generation. Microgrids within substations enhance system resilience, enable energy sharing, and improve power quality.

Cybersecurity and Grid Resilience:

With increasing digitization and connectivity, cybersecurity is of paramount importance for substations. Advanced cybersecurity measures, such as encryption, intrusion detection systems, and secure communication protocols, are anticipated to be implemented within substations to protect against cyber threats.

Enhanced grid resilience features, such as rapid fault isolation and self-healing capabilities, will be integrated into substation designs to minimize disruptions caused by cyber attacks or physical events.

Environmental Sustainability:

Substations will continue to adopt environmentally friendly practices to minimize their ecological footprint. Efforts will be made to reduce greenhouse gas emissions through the use of eco-friendly insulating gases, such as sulfur hexafluoride ($SF6$) alternatives, and the integration of renewable energy sources within substations. Energy-efficient designs, energy management systems, and sustainable construction practices will be embraced to enhance the environmental performance of substations.

Conclusion:

Anticipated developments in substation technology and infrastructure are poised to revolutionize the power industry. Digitalization, intelligent substations, FACTS devices, HVDC, power electronics, energy storage systems, grid-forming inverters, microgrids, cybersecurity measures, and environmental sustainability will shape the future of substations. These advancements will improve substation performance, grid reliability, and energy efficiency while enabling the integration of renewable energy sources and supporting a more sustainable and resilient power system. As the power industry continues to evolve, substations will play a crucial role in realizing a cleaner, more efficient, and reliable energy future.

Challenges and Opportunities in the Evolving Energy Landscape

The energy landscape is undergoing a profound transformation driven by various factors such as the need for decarbonization, increased renewable energy integration, advancements in technology, and changing consumer demands. This section elaborates on the challenges and opportunities presented by the evolving energy landscape.

Challenges:

Decarbonization and Climate Change Mitigation:

- The transition to a low-carbon economy presents challenges in reducing greenhouse gas emissions and mitigating the impacts of climate change.
- The integration of intermittent renewable energy sources into the grid requires addressing the challenges of intermittency, storage, and grid stability.

Grid Integration of Renewable Energy:

Integrating high levels of renewable energy into the grid poses challenges related to grid stability, transmission capacity, and balancing supply and demand. The intermittent nature of renewable sources requires the development of sophisticated grid management systems to ensure reliable and efficient power delivery.

Aging Infrastructure:

Many power systems and electrical grids globally are facing aging infrastructure, which requires significant investments in refurbishment, modernization, and grid resilience.

Upgrading infrastructure while ensuring uninterrupted power supply is a challenge that utilities and grid operators must address.

Cybersecurity and Data Privacy:

The increasing digitalization and connectivity of energy systems introduce cybersecurity vulnerabilities and the risk of data breaches. Protecting critical infrastructure and ensuring the privacy and security of consumer data are challenges that require robust cybersecurity measures and regulations.

Evolving Regulatory Frameworks:

The evolving energy landscape necessitates adapting regulatory frameworks to accommodate new technologies, business models, and market structures.

Striking a balance between promoting innovation, maintaining grid reliability, and protecting consumer interests poses challenges for regulatory bodies.

Opportunities:

Renewable Energy Development:

- The transition to renewable energy sources presents significant opportunities for expanding clean energy generation.
- Advancements in solar, wind, and energy storage technologies offer opportunities for increased renewable energy deployment and cost reductions.

Energy Efficiency and Demand Response:

Improving energy efficiency in buildings, industry, and transportation sectors presents opportunities for reducing energy consumption and greenhouse gas emissions.

Demand response programs, enabled by smart grid technologies, allow consumers to actively manage their energy usage and contribute to grid stability.

Electrification of Transportation:

The shift towards electric vehicles (EVs) offers opportunities for increased electricity demand and the integration of EV charging infrastructure.

Smart charging solutions and vehicle-to-grid (V2G) technologies provide opportunities for bidirectional energy flow, grid support, and load management.

Digitalization and Advanced Grid Technologies:

Digitalization and advanced grid technologies, such as smart meters, IoT-enabled devices, and data analytics, offer opportunities for improved grid monitoring, control, and efficiency.

Advanced grid management systems, including AI-based algorithms and predictive analytics, enable optimized power flow, grid stability, and demand response.

Distributed Energy Resources and Microgrids:

The proliferation of distributed energy resources (DERs) such as rooftop solar panels, energy storage systems, and microgrids presents opportunities for decentralized and resilient energy systems.

Localized energy generation, coupled with smart grid technologies, enables increased energy independence, grid resilience, and community empowerment.

Energy Market Transformation:

Evolving energy markets, including peer-to-peer energy trading and community energy projects, offer opportunities for consumer participation, energy democratization, and new business models.

Prosumers (consumer-producers) can actively participate in energy markets by generating, storing, and trading their excess energy.

Conclusion:

The evolving energy landscape presents both challenges and opportunities for the power sector. Overcoming challenges such as decarbonization, grid integration of renewables, aging infrastructure, cybersecurity, and evolving regulatory frameworks requires collaborative efforts among stakeholders, policymakers, and industry players. However, the opportunities for renewable energy development, energy efficiency, electrification of transportation, digitalization, and distributed energy resources offer prospects for a cleaner, more efficient, and sustainable energy future. By embracing innovation, adopting new technologies, and promoting a resilient and flexible energy system, the evolving energy landscape can drive the transition towards a more sustainable and prosperous future.

Concluding Remarks on the Significance of Electrical Substations

Electrical substations play a pivotal role in the reliable, efficient, and safe delivery of electricity to homes, businesses, and industries. Throughout this book, we have explored the various aspects of electrical substations, including their historical background, components, design considerations, safety protocols, environmental impact, and future outlook. In these concluding remarks, we emphasize the significance of electrical substations and their contribution to the power sector.

Backbone of Power Distribution:

Electrical substations serve as the backbone of power distribution systems, enabling the transmission and transformation of electricity from high-voltage transmission lines to lower voltages suitable for distribution. They facilitate the efficient transfer of electrical energy, ensuring its reliable delivery to end-users.

Grid Reliability and Stability:

Substations are vital for maintaining the reliability and stability of the power grid. They incorporate protective devices, control systems, and monitoring equipment that ensure the safe and continuous operation of the grid. By regulating voltage levels, managing power flow, and coordinating system protection, substations help prevent power outages, reduce downtime, and enhance grid resilience.

Integration of Renewable Energy:

As the world increasingly transitions towards renewable energy sources, electrical substations play a crucial role in integrating these intermittent sources into the grid. They facilitate the connection of renewable energy generation, such as solar and wind farms, to the power system, enabling the efficient utilization and integration of clean energy.

Grid Flexibility and Management:

Substations, equipped with advanced control systems and communication technologies, enable grid flexibility and management. They support load balancing, reactive power control, and voltage regulation, ensuring efficient utilization of generation resources and maintaining grid stability. Substations also enable grid management practices like demand response, grid automation, and optimization of power flow.

Safety and Reliability:

Safety is of utmost importance in electrical substations. Strict adherence to safety protocols, equipment standards, and personnel training ensures the well-being of substation workers and the public. Substations incorporate protective relays, circuit breakers, and grounding systems to safeguard personnel, equipment, and the surrounding environment.

Environmental Responsibility:

Electrical substations are increasingly embracing sustainable practices to minimize their environmental impact. Efforts are made to reduce greenhouse gas emissions, optimize energy consumption, and incorporate renewable energy sources. By implementing eco-friendly designs, adopting energy-efficient technologies, and promoting sustainable construction and operations, substations contribute to a greener and more sustainable energy infrastructure.

Technological Advancements and Innovation:

Substations are at the forefront of technological advancements in the power sector. Digitalization, intelligent systems, advanced monitoring and control technologies, and grid automation are transforming substations into smart entities. These innovations enhance the efficiency, reliability, and performance of substations, allowing for real-time monitoring, predictive maintenance, and optimized grid operations.

In conclusion, electrical substations are the vital link in the power supply chain, enabling the transmission, transformation, and distribution of electricity. They ensure grid reliability, integrate renewable energy sources, support grid flexibility, and prioritize safety and environmental responsibility. As the energy landscape evolves, substations will continue to evolve, incorporating advanced technologies, meeting regulatory requirements, and embracing sustainable practices. The significance of electrical substations in delivering a reliable, sustainable, and resilient power supply cannot be overstated. It is through the continuous advancement and improvement of electrical substations that we can build a more efficient, cleaner, and sustainable energy future for generations to come.

APPENDIX: GLOSSARY OF KEY TERMS

This glossary provides definitions and explanations of technical terms used throughout the book on electrical substations.

Electrical Substation: An electrical substation is a part of the power system that transforms voltage levels, regulates power flow, and facilitates the distribution of electricity from high-voltage transmission lines to lower voltage levels suitable for distribution to end-users.

Transformer: A transformer is a device used in substations to transfer electrical energy between two or more electrical circuits through electromagnetic induction. It steps up or steps down voltage levels to ensure efficient transmission and distribution of power.

Circuit Breaker: A circuit breaker is an electrical switch designed to protect electrical circuits from damage caused by overcurrent, short circuits, or other faults. It interrupts the flow of electric current when a fault is detected to prevent further damage.

Switchgear: Switchgear refers to the combination of electrical disconnect switches, circuit breakers, and protective relays used to control, protect, and isolate electrical equipment in substations. It enables safe operation, maintenance, and troubleshooting of electrical systems.

Busbars: Busbars are conductive metal bars used to carry and distribute electric power within substations. They provide a common connection point for electrical equipment and facilitate the transfer of electrical energy between different components.

Protection Relays: Protection relays are devices that monitor electrical systems and detect abnormal conditions or faults. They initiate protective actions, such as tripping circuit breakers, to isolate faulty equipment and protect the power system from damage.

Capacitors: Capacitors are devices used in substations to store electrical energy. They provide reactive power support, voltage regulation, and power factor correction, improving the efficiency and performance of the power system.

Reactors: Reactors are electrical devices used in substations to control current flow, limit short-circuit currents, and provide reactive power compensation. They help maintain system stability and reduce the impact of transient disturbances.

SCADA (Supervisory Control and Data Acquisition) System: SCADA systems are computer-based control systems that monitor and control substations and power systems. They provide real-time data, remote control capabilities, and data acquisition for efficient and reliable operation of substations.

Protection Scheme: A protection scheme refers to a coordinated set of protection relays and devices designed to detect faults, isolate faulty equipment, and ensure the safety and reliability of the power system. It comprises various protection functions, such as overcurrent, differential, and distance protection.

HVDC (High Voltage Direct Current): HVDC is a technology used for transmitting high-voltage direct current over long distances. It allows for efficient power transmission, interconnection of different grids, and integration of renewable energy sources.

Microgrid: A microgrid is a localized power system that can operate independently or connected to the main grid. It integrates distributed energy resources, energy storage, and smart grid technologies to provide reliable and resilient power supply to a specific area or community.

Grid Codes: Grid codes are technical standards and regulations that specify the requirements for the connection and operation of power generation, transmission, and distribution systems. They ensure the reliable and safe operation of the grid, maintain power quality, and promote interoperability.

Energy Storage Systems (ESS): Energy storage systems store electrical energy for later use. They play a crucial role in smoothing out fluctuations in renewable energy generation, providing backup power, and supporting grid stability and resilience.

Intelligent Substations: Intelligent substations incorporate advanced control systems, sensors, and communication technologies to enable real-time monitoring, diagnostics, and optimization of substation operations. They enhance the efficiency, reliability, and safety of substations.

Grid Resilience: Grid resilience refers to the ability of the power system to withstand and recover from disruptions, such as extreme weather events, natural disasters, or cyber-attacks. Resilient grids incorporate measures to quickly detect, isolate, and restore power to minimize downtime and maintain critical services.

Demand Response: Demand response refers to the active participation of consumers in adjusting their electricity usage in response to signals or incentives. It helps balance supply and demand, manage peak loads, and optimize grid operations.

Prosumers: Prosumers are consumers who also produce electricity, typically through the use of distributed energy resources (DERs) like solar panels or wind turbines. They can contribute excess energy back to the grid or participate in energy trading.

www.ingramcontent.com/pod-product-compliance
Lightning Source LLC
Chambersburg PA
CBHW072212290526
45794CB00004B/1730